A PICTURE BOOK FOR PALEONTOLOGISTS

HOW FAULTED IMAGES AND METHODS HAVE SUBVERTED THE STUDY OF HUMAN ANCESTORS

Jon Bogle

A PICTURE BOOK FOR PALEONTOLOGISTS

HOW FAULTED IMAGES AND METHODS HAVE SUBVERTED THE STUDY OF HUMAN ANCESTORS

Jon Bogle

Figurative images in paleontological books, text books and museum dioramas have the same set of incorrect ideas that beginning art students render. These faulted images have subverted paleontological analysis and the scholarly text. A goal of this book is to show how to accurately compare two skulls from two different primate species without the distortions caused by faulted imagery.

SCI 054000 Science Paleontology

ART 009000 Criticism Theory

PHO 013000 Plants Animals

JON BOGLE

Professor Emeritus: Lycoming College,
Williamsport, Pennsylvania, 1976-2002

Deborah, to whom I owe all

Acknowledgements:

David Brill, whose stunning photographs of human skulls provided the information needed for the comparisons.

My great thanks to Professor Colin P. Groves, University of Canberra, for generously forwarding prints of The Gorilla Growth Series.

The late John Larkin, volunteered his considerable knowledge of grammar to do an earlier edit. His precise, generous, and comprehensive work was critical.

Deborah, who sat with me doing a final line by line edit. She turned up 300 to 400 corrections in a manuscript I had thought was nearly perfect.

Professor Leedom Lefferts, who generously gave a thorough review and offered great council on clarifying the intent of the book's layout and the examples used.

SUMMARY

This is a work of observation rather than scholarship. The numerous photos and drawings illustrate the observations, the text explains them.

What I have to offer is my training in modeling and rendering the classical naturalistic figure. Scholarly paleontology is hamstrung by a set of visual image fallacies which are universal and are the same ones that beginning art students draw.

Beyond faulted imagery, paleontologists have an arcane scholarly methodology that is more like monastic studies than science. Paleontologists study the evolution of physical traits trying to find a sequential line between ancient forms and modern ones. Traits aren't connected to a sequential lineage, but instead linked to function. Traits change when function changes. Evolutionists are locked into Natural Selection as the creator of new species but while it picks the winners and losers, it doesn't cause the changes that lead to new species.

Life is self programing. Its history shows that it was very slow to gain complexity but when conditions and abilities finally came together, it unleashed an astonishing explosion of diversity. Evolution lives in the present, doesn't know or plan for the future, and sometimes produces bad design. It can only modify what already exists so its early formulations sometimes force it into compromised solutions.

Animal heads are one example of particularly bad design. Heads are a dangerous compromise made necessary because early animals articulated their chemical messengers into nerves. Nerves still depend on a chemical transfer to function, making their transmission rate quite slow.

Heads are the result. The central processor is placed in the front where it monitors air, water, and food intake. The vital senses are plugged directly into the brain and positioned so the head moving forward can read its surroundings. Land vertebrates follow strict ancient protocols for how they face into the environment. Bioengineering protocols embedded deeply in the genetic code maintain essential relationships that are needed to sustain life.

Genetics are structured like an onion. The inner core exerts controls and regulates essential processes. The core is shared with most other living organisms. Only the outside layers can be modified. It is only these surface layers that differentiates us from an elephant. Ancient protocols modified relationships within our skulls so we can face correctly into the environment like other animals. Our bipedal stance produces almost all the differences with our ape relative's skulls. The ancient protocols tripped a reset when we stood up.

TABLE OF CONTENTS

A PICTURE BOOK FOR PALEONTOLOGISTS

As a sculptor, I am an outsider in this discussion. I was taught the classical human figure at Temple University's Tyler School of Art which, at the time, was a bit anachronistic. My first undergraduate years were spent in the studio with live models, modeling the figure in clay, drawing the figure, and painting the figure. While I seldom represent the figure in my own art work I found this foundational training to be invaluable. When I took a position at Lycoming College, I instituted a figure modeling course which I taught every semester for twenty-five years.

While trying to teach my students how to see the figure, I found I was forced to work re-actively against a set of their embedded assumptions. These assumptions were, I eventually discovered, an expression of a universal human developmental stage. My students thought about the figure in the same way it has been depicted in traditional cultures from around the world and spanning thousands of years.

The figurative imagery in anthropological articles and books has the same set of incorrect assumptions that my beginning students insisted on using. The analysis in the text is then based on the same incorrect figurative assumptions as the illustrations. The issues I am raising are not new information. Greek sculptors working twenty-three hundred years ago had a much better understanding of the human figure, its proportions, its movement, its balance, and its physiology, than paleontologists have today.

Rendering the naturalistic figure is similar performing an Olympic high dive. Anyone can see the dive but unless you have spent years of training, working with an experienced coach, doing the dive correctly is highly unlikely. Rendering the figure is a performance just as competent diving is a performance. Ancient Greek sculptors needed two hundred years to learn how to do the naturalistic figure. This experiential knowledge requires master sculptors to coach their apprentices. This knowledge transfer was lost when the Roman empire collapsed. In the time leading up to the Renaissance, Italian artists required another two hundred years to relearn it.

SCHEMATIC FIGURE

To grasp why understanding the figure is such a problem we have to look at the structure of perception. All sensations coming into our minds are analog. Light and sound waves are read as sight and hearing, direct readings of chemicals are understood as taste and smell, and pressure against our skin gives tactile awareness. Our minds can't store these cumbersome analog events but convert them into greatly simplified schematics. This is very efficient because our simple mental schematic for tables, trees, or persons will quickly

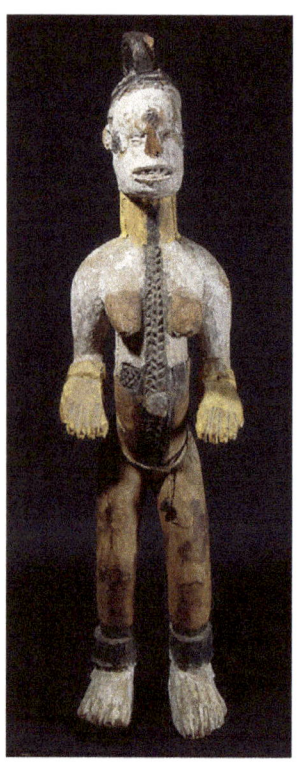

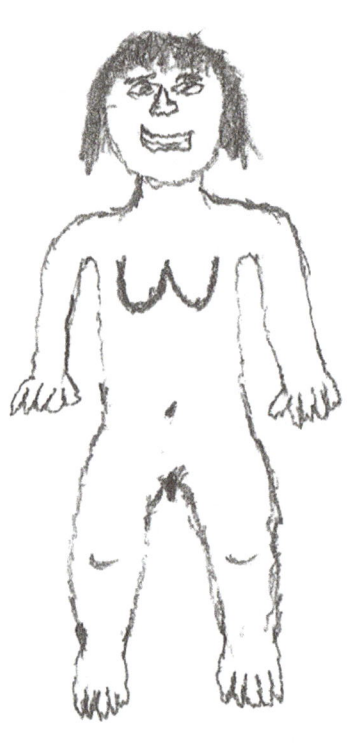

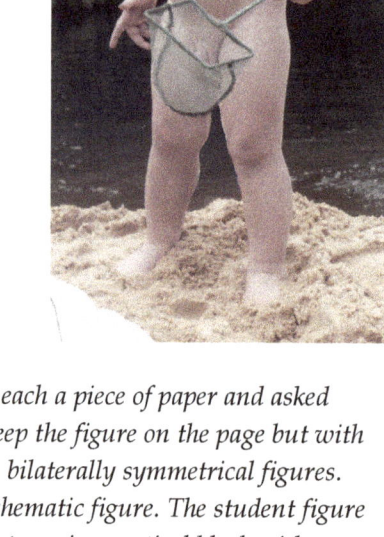

Igbo female shrine figure (Alusi) photo: Michael Cichon ©Michael Cichon Tribal Arts

On the first day of drawing class I gave the students each a piece of paper and asked them to draw a figure. The only instruction was to keep the figure on the page but with only three exceptions they all drew standing, frontal, bilaterally symmetrical figures. With no model to refer to, they drew their internal schematic figure. The student figure has large a head set directly on top of a thin neck, the torso is a vertical block with arms out to the side with short legs coming straight down. The eyes are high in the head.

The description of the tribal figure exactly fits the student's except that the eyes are not set high in the head. I suspect the tribal sculptors who do these figures also carve masks out of which the wearer must see. Note the emphasis in both the tribal and student figure to show all the fingers and toes.

The photo of the child is strikingly similar in proportion and attributes to the student's drawing. A description of his physical attributes would match the description of the other two images.

identify all possible tables and trees and all possible persons from all possible viewpoints. When a new student or tribal artist is asked to draw or carve a figure what they render is not how the figure looks, but their mental schematic image for it. The universal schematic figure has the proportions and attributes of a toddler. An image which we all learned by tactile exploration during that stage of our lives.

The Ancient Greeks, and later the Italians, reprogrammed their internal figurative image from that of a child to that of an adult. Interestingly, when children were portrayed in ancient Rome they often were given adult proportions. It took more observation to be able to see the child again. The naturalistic figure is external information while the schematic child image is internal. Artists use nude models to challenge their internal image and, if successful, they are able to replace it with an elaborate program for understanding the naturalistic figure.

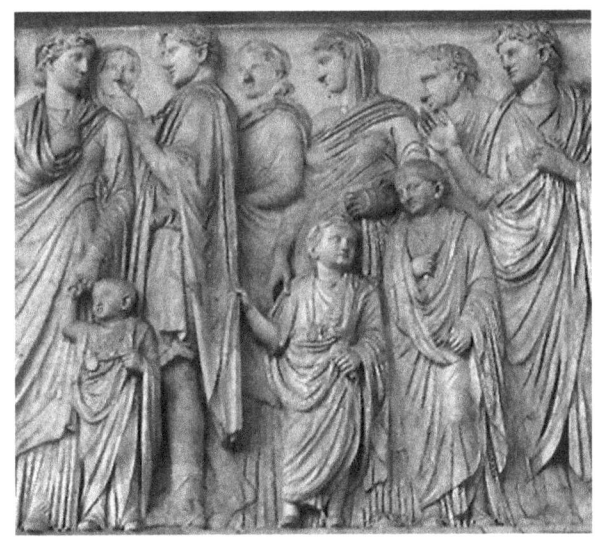

Note that the three children in front especially the two farther out are like small adults while the one in the middle has more childlike proportions.

Ara Pacus, Rome, 9 bce Wikipedia Creative Commons Attribution-Share Alike 3.0

TABLE FALLACY

The other type of faulted imagery in paleontology is the Table Fallacy. Gravity enters our creative processes and effects our intellectual and artistic expression. The conceptual pull of gravity often makes itself felt in the works of painters, even in paintings of abstract and nonrepresentational intent. When teaching, I would sometimes employ a standard teacher's device of turning the canvas or sculpture upside down or sideways. This exposes all the elements of the design constrained by the emotional need to rest shapes on a base for support. To break free from the pull of gravity sculptors rely on armatures. Armatures are temporary structures made of wire and wood that support wet clay. When my students used a board or table as a base on which to model clay, their creations would spread out along the board's surface like a fungal growth. I would then help them to construct an armature so that their ideas could lift up off the table and move more freely in space. I call this imposition that gravity has on thought: the Table Fallacy.

The "Table Fallacy" is rampant in the scientific study of heads and comes from the convenient position that skulls are placed in as they are studied. Any skull, no matter the animal, will become stable if it is placed so its jaw sits flat on the table and a prop is put under the skull case to support it. This arrangement creates a stable triangular support. For generations of scientists, this position of convenience has been the standard position in which skulls are placed when they are compared or illustrated.

This image of the skull sitting propped on a table has subversively imposed itself on scientists' understanding of skulls. In the literature of evolutionary anatomy, and in the illustrations used to support these writings, there is an unconscious transfer from the position of the skull on the table to assumptions of how the living animal held its head. Text book skulls are almost invariably depicted in the Table Fallacy position. Even when the animal is fleshed out, with skin and fur, it still stays in the Table Fallacy position.

Front and side views of one of the skulls in Uno Samuelson's Gorilla Growth Series photographed in the Table Fallacy position. In the frontal view the eyes have rotated upwards as they do in untutored drawings. The side view shows prognathism, the projection of the face, which doesn't exist except as a artifact of the Table Fallacy

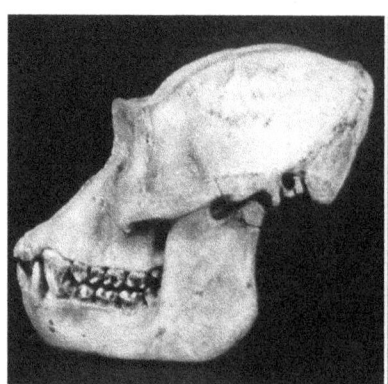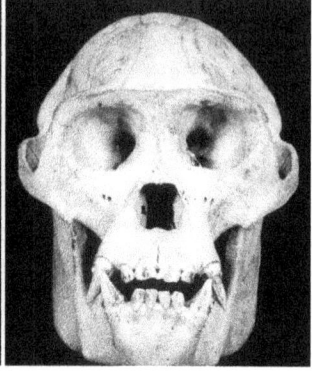

Eyes on Top

The comparisons between skulls of different species are badly skewed. The first rule that artists teach their students about the human head is that the eyes are always located on a horizontal axis halfway between the top of the head and the chin. This, for the human head, is followed by the rule of diminishing halves: the tip of the nose is halfway between the eyes and the chin, and the mouth is half way between the bottom of the nose and the chin. These last two proportions are not very solid and vary a bit from person to person, but they are helpful for getting the student into the ball park.

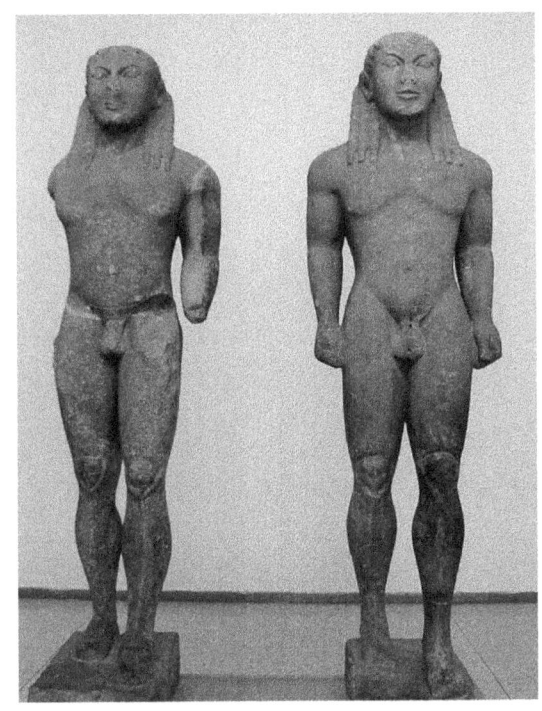

Kleobis and Biton by Polymedes of Argos. Marble, ca. 580 BC. 6 ft. 5 ½ in. Archaeological Museum of Delphi, Wikimedia Commons Adam Carr

These figures, although moving towards naturalism, share many aspects of the student's figures including high eyes.

I have taught hundreds of students over the years to model portrait heads in clay, and spent a good portion of that time tearing out and relocating eyes. This is not only a proclivity of my students; it is common whenever the figure is thought of in concept rather than as a visual object. Traditional and tribal artists, the earliest Greek sculptors, and the ancient Babylonians could all have used my friendly eye relocation services.

We are programed to look intensely at faces. When we first draw we will usually draw just the face and use that image to depict the entire head. Eyes are in the upper third of the face so these first, untrained drawings have the eyes towards the top of the head. Artists needed a simple rule so their students could locate the eyes in the center of the head and recognize the part of the head that is above the hairline.

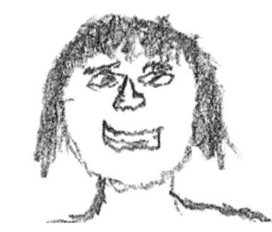

College student head drawing from memory. (Schematic figure)

A breakthrough occurred when I realized that the eyes in the middle rule that I had been taught and taught my students worked as an effective guideline for other animals. It is nor only applicable to other primates, but is also relevant to cats, dogs, and all other vertebrates.

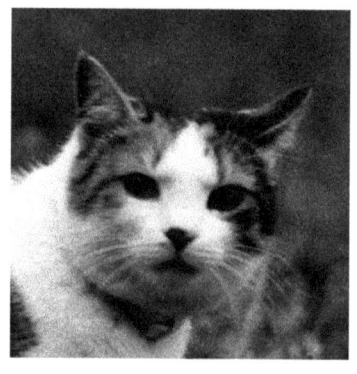

Tryout the Table Fallacy. Put your thumbs under your jaw line and push your head up until the base line of the jaw is hori-

Patricia, the family house cat. Note, eyes are halfway from top to bottom.

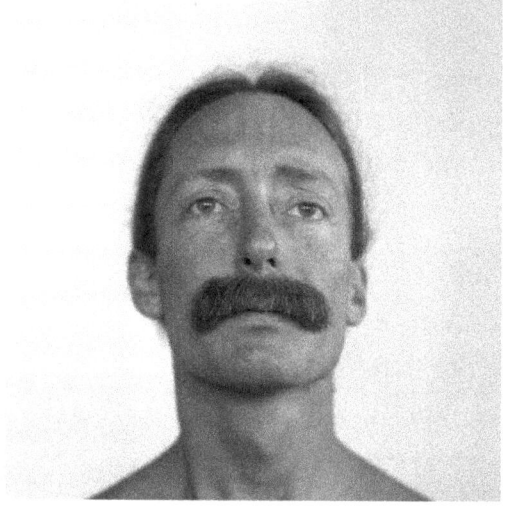
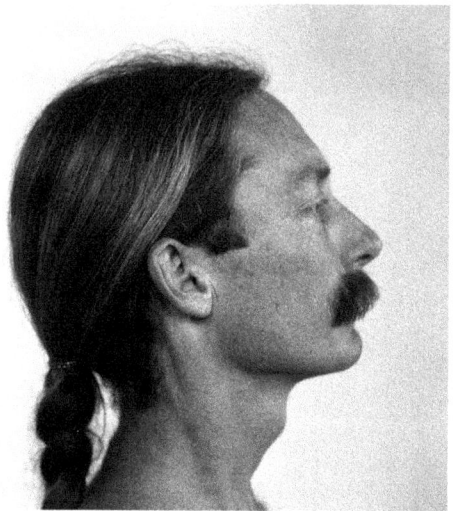

Here, the model is posed in the Table Fallacy position with the eyes rotated upward. From the side view the mastoid muscles incline forward from behind the ear to the base of the throat which is how they are illustrated in Aiello and Dean's "Human Evolutionary Anatomy" pp 223 224 figs 12-17 12-19. Below is the model posed naturally where the mastoids descend vertically as the neck inclines forward to put the face in front of the body.

zontal. This, in living human terms, is the Table Fallacy. How are you doing? You have lost some of the overlap of your binocular vision because your nose is now getting in the way. The snout is evident in skulls placed in the Table Fallacy position and is used as evidence, in numerous scientific papers, that the creature had limited binocular vision. The Table Fallacy inclines your face thus increasing the length of your snout, your degree of prognathism. Prognathism is thought to be a regressive trait so I wouldn't stay there too long. It could be much worse: think of the difficulties of a baboon forced to hold its head in the Table Fallacy position.

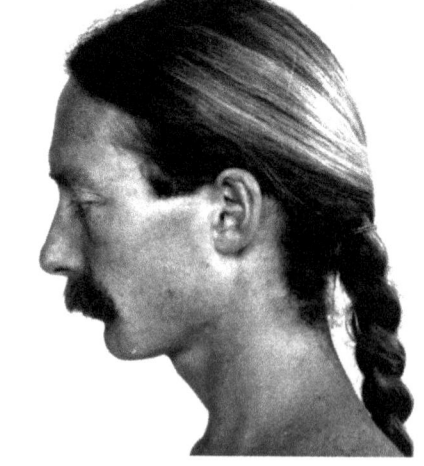

Osman Hill

Osman Hill was concerned about the problem he thought baboons must be having when he wrote:

> *"Possession of powerful teeth subsumes a certain forward projection of the facial structure in which they are implanted (prognathism) and also the bulky muscles of mastication to assist the jaws in their work....Such imbalance is of an even greater degree in the heavily jawed baboons; in these as in the Gorilla, there is an enormous expansion of the nuchal area for the accommodation of the dorsal neck muscles "*
>
> (Evolutionary Biology of the Primates Osman Hill, Academic Press, London New York 1972, pp. 72-73).

Hill, from studying the skull in the Table Fallacy position, found a need to support a forward projecting snout. In the quote, Osman Hill speculates that the expanded nuchal crest area in the rear of the baboon's skull has the function of attaching to the large dorsal muscles whose function would be required to balance a head in the Table Fallacy position.

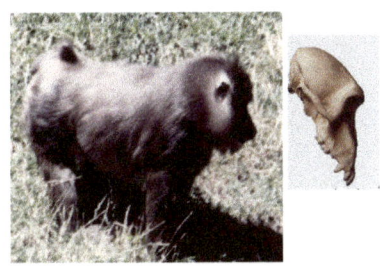 Hill needn't have worried about baboons in this regard. Both we and baboons place our faces in the environment in the same sequence and manner dictated by the sense organs of the head. In both cases, the face is situated below the brain case. In man, the brain case sits like an egg on a table with face growing downward from below the small front end of the egg.

Baboons, who approach life with their spinal columns held horizontally have rolled their brain case forward until it is almost vertical above the face, more like an egg balancing on its small end. The eyes are at the center in this changing configuration between man and monkey, always remaining in the middle of the head's image. In all primates, humans included, the spine meets the floor of the brain case at an angle that is obtuse in front and acute to the rear.

The mechanism for a baboon turning its head is quite different from that in humans. To look left and right, we swivel our heads by rotating the neck vertebra on their long axis. Our heads pivot on top of our more vertical spinal columns. The baboon, with a horizontal spinal column, looks left and right by curving the neck. The muscles that turn the baboon's head to look left and right attach to the nuchal area on the back of the skull.

The expanded area that Osman Hill commented on results from the dorsal muscles being the primary controllers of head motion in baboons. To look left and right, baboons have an enlarged nuchal process and dorsal muscles; the same muscles we use when we lean our heads over to touch our ear to our shoulder. The muscles that move our head as we pivot it to look left and right are the mastoid muscles, the two strong cord-like forms that start just behind the ears and descend vertically to meet at the base of the throat. In humans, the mastoids turn the head to align sight.

STUDENT'S SCHEMATIC FIGURE

On the first day of a spring semester drawing course, I gave each student a sheet of paper with written instructions: "Draw a (male) (female) nude complete figure in the space below." I was interested in seeing the figure that the students would draw before work in the drawing course would alter it. Drawing without a model would reveal, I hoped, the student's internal figure.

A number of students in this class had just completed my fall figure modeling course and had considerable experience in rendering the human figure. I was surprised and a bit taken aback by the formulaic drawings from these students.

On further study, this group revealed some interesting connections to the development of the realistic figure in historic times. While the beginning students' figures have a toddler's figurative proportions, these second semester students drew their figures proportions more like that of an adolescent.

Their body proportions are more consistent with those in Archaic Greek figures whose sculptors, like these students, were in an early phase of learning to render the figure accurately. Like the beginning students, these returning student figures were relatively consistent with each other and were more likely to show some motion or to break from bilateral symmetry. The first-time student figures are more consistent with many tribal figures where no attempt at naturalism is made.

Half the male students were asked to draw a female figure and half the female students were assigned a male figure. My intent was to see if I could identify sex-based differences in the images, but I wasn't able to find any. In the drawings being presented, I have indicated the gender of the person doing the drawing.

I asked, "Have you ever had a figure drawing or a figure modeling course before? Yes-No circle one." This answer is also written beneath each figure. The students who answered "yes" were predominately the students from the figure modeling course the semester before. One student who answered "no" may not have actually had a figure course but was a transferring art major with apparent drawing experience. He also drew a bilaterally static figure, but it was the only back view and it had more natural proportions.

The figure, in concept, is complete with all its parts. Some of the beginning students even took pains to draw all ten fingers and all ten toes. These drawings closely follow a thought that the torso as a block, beneath which are supporting legs with arms attached to its outside.

Directly after the students finished drawing the first figures, I posed a model and had a normal first day drawing session for the remaining two hour class. In a following class, two days later, after a couple of short warm-up poses with the model, I again gave the students paper and had them draw the model for about ten minutes. The results are the second figures on the right of each set. During both drawing sessions in which these works were produced I offered no instruction or help. Less than two hours of instruction separates the two sets of drawing.

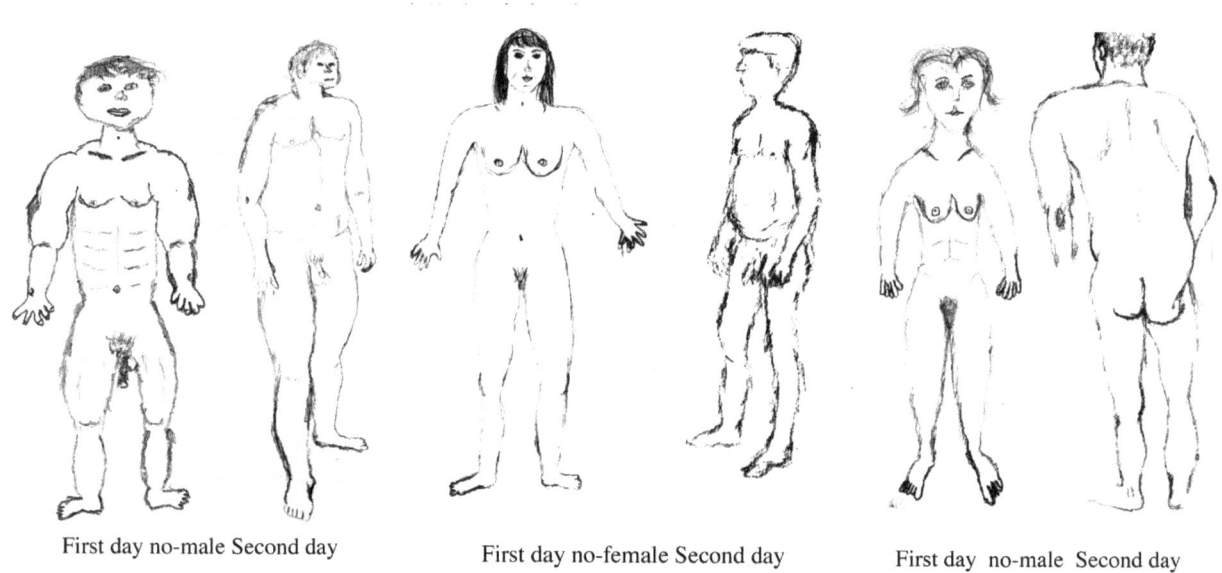

First day no-male Second day   First day no-female Second day   First day no-male Second day

The obvious differences between the two sets are not the result of much training or in-creased skill but rather how they approach the task. In the first set, the students were asked to draw a figure from memory and they, like tribal artists, used their schematic image for a person. On the second day, they were asked to draw from a model using the external visual image as a resource. The intervening hour and a half of instruction had primarily taught them how to approach this task.

The second set of drawings is much more naturalistic than the first ones, but are compos ites made from what the student saw and what they thought. Learning to draw realistically requires artists to defeat their preconceived ideas. These drawings are only the first halting step in that process.

The major attributes of the schematic figure that distinguishes it from the actual human form are its lack of movement, rigid verticality, strict bilateral symmetry, and a set of proportional differences in the legs, arms, head, and body.. The conceptual figure is made up of its parts and is constructed by adding these units together.

The conceptual figure in sculpture or drawing closely follows a verbal description of the figure. Described, the figure has a head, a trunk, two arms, two legs, two hands each with five fingers, two feet each with five toes and so on. Of course, the human figure has these things too, but in the human figure the elements flow together to make an organic whole.

My most difficult task in teaching the figure or the portrait head is to have the students understand that the actual human form is a single whole instead of a collection of parts. The parts of the figure for which we have names: shoulders, knees, noses, flow into other elements and across large areas of transition. The student's work assembles the elements of the figure like a Lego construction. The work is either assembled from elements or, in the case of the head, the elements, such as the nose and ears, are applied to a basic, ball-like form.

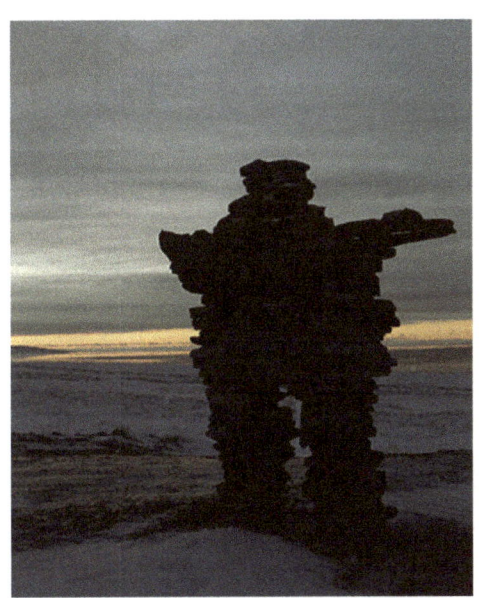

Wikimedia Commons
photo: Nicolas M. Parrault

Inuit Inuksuk rock-piled sculpture and my young daughter's ceramic sculpture.

Both figures demonstrate how little specific information is needed to convey the image of a person.

I have mistaken a fence post at a distance for a person. My dog became very defensive when he suddenly noticed my jacket thrown on another post which made a person-like image. Hackles up, growing, backing off. "Who are you? When did you get here? Boss, do something."

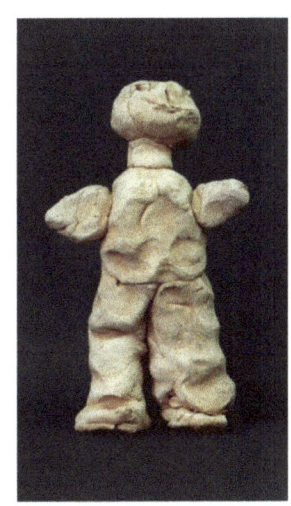

GOOD POSTURE

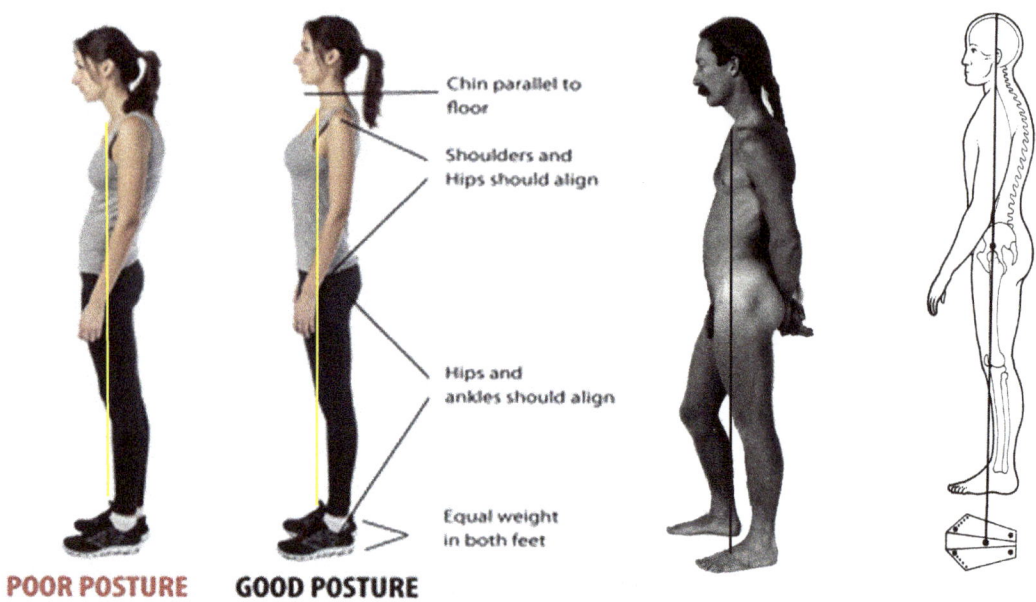

Chin parallel to floor

Shoulders and Hips should align

Hips and ankles should align

Equal weight in both feet

POOR POSTURE   **GOOD POSTURE**

Going to the internet and searching for images of good standing posture will turn up hundreds of the type of illustration on the left showing good and bad posture. The pose marked as bad is a slouchy version of a natural stance while the good posture version is completely artificial. The supposed good posture is an attempt to achieve the schematic figure in a living pose. Like the soldier at attention, the girl showing "good posture", and the Sun God, the illustration on the right is based on the schematic figure. The far right figure is an illustration from Aiello and Dean's " Human Evolutionary Anatomy," page 246 Figure 14-3, which depicts the figure's center of gravity. It is correct in just two places. The center passes through the ear in a naturally standing figure and lands near the arch of the supporting foot. Both these points are correct in the illustration but every thing else is in-correct. It assumes that the center of gravity would be in the middle of the figure and pass through the center of the hip, the knee, and the shoulder. To make this happen the base of the spine has been moved forward to an impossible place.

The naturally standing model inclines the neck forward to place the face in front of the body just like all land vertebrates. A simple observation is to look at the mastoid muscles, the strong cords that start just behind the ears and descend to the base of the throat. In a naturally standing figure the mastoid descends vertically but in the schematic based figure they incline forward. (see photos and discussion on page 6)

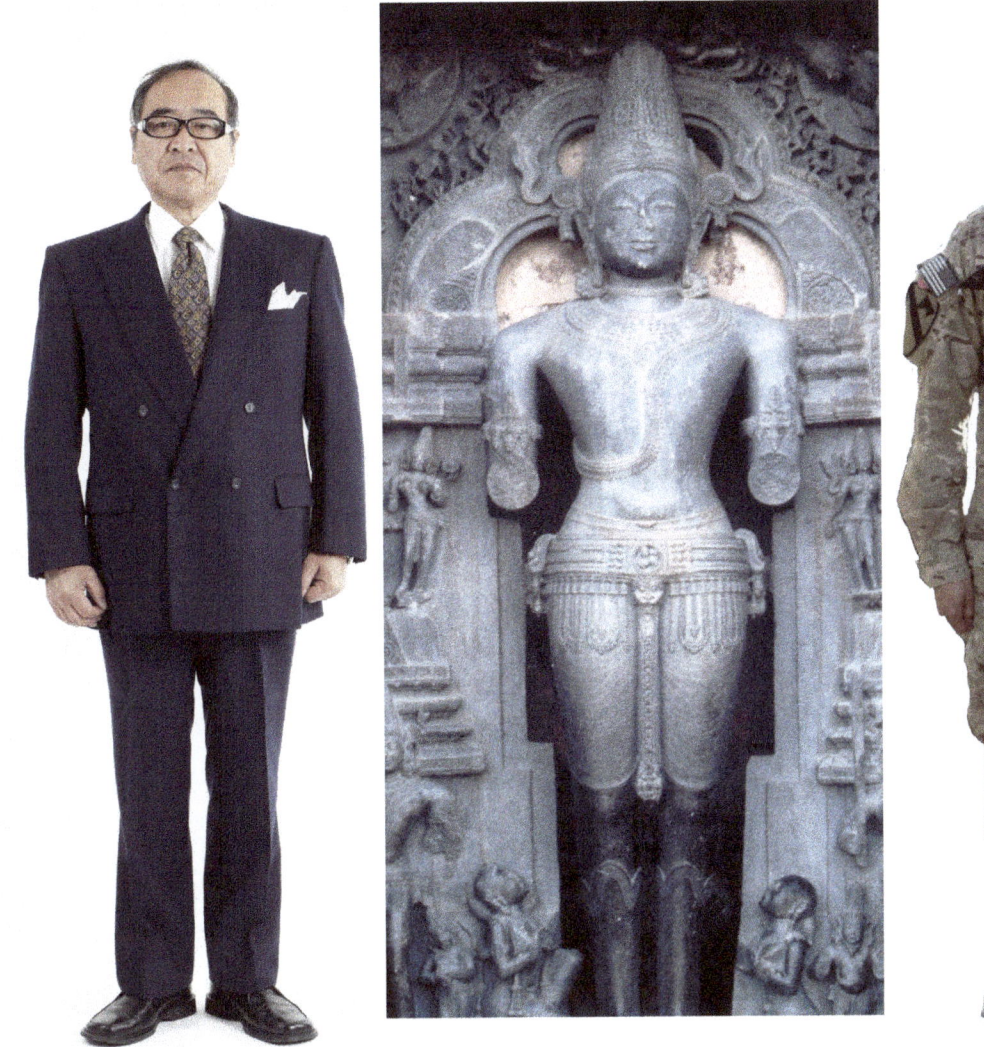

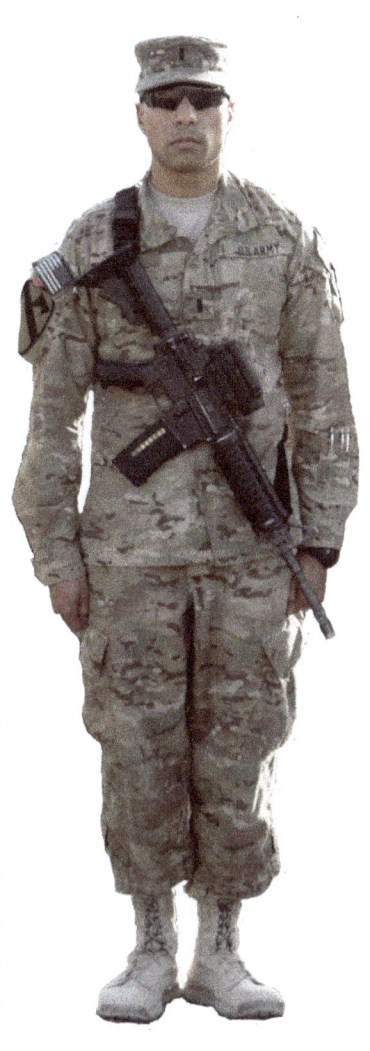

The schematic figure is also the universal power figure in a way that the naturalistic figure is not. The man in a suit and the soldier use stance and clothing to visually recreate the schematic power figure thereby giving themselves added authority. The business suit's coat lengthens the torso and renders it block like. It also shortens the legs and makes the arms appear to be attached outside the torso. The soldier's fatigue jacket does the same thing. The figure in the center is a twelfth century Sun God from Konarak in eastern India. The Sun God is also rigidly vertical, has short legs, arms that attach to the outside of a torso block without movement. Unlike the God, however, the small ancillary figures in the same panel are expressive and move. It is primarily the vertical rigidity that denotes the God and the schematic student figures on the previous pages. In traditional cultures, across time and religions, figures of the gods are depicted with these same attributes.

It seems more than a bit unfair that business women must compete with their male counterparts who are all wearing the God-like power costume.

14

SKULL DESIGN

All land vertebrate heads are aspects of the same head design program. This includes the great apes, primates, mammals and finally all vertebrates. Evolution is a conservative process and it keeps working configurations stable unless they are challenged by reality. When challenged, evolution can only modify existing structures, like turning scales into feathers and hair or suppressing others like hens teeth and fur on humans.

We are trying to understand the engineering design system for heads. This can be revealed by finding the elements of the design which are stable, where the design has places of flexibility and elements of the structure that serve a function. Complex organizations are more easily analyzed by first determining the organization's function. Then finding elements in its structure are fixed and immutable and where elements can be modified. This is in essence how the ancient Greeks worked out how to render the naturalistic figure.

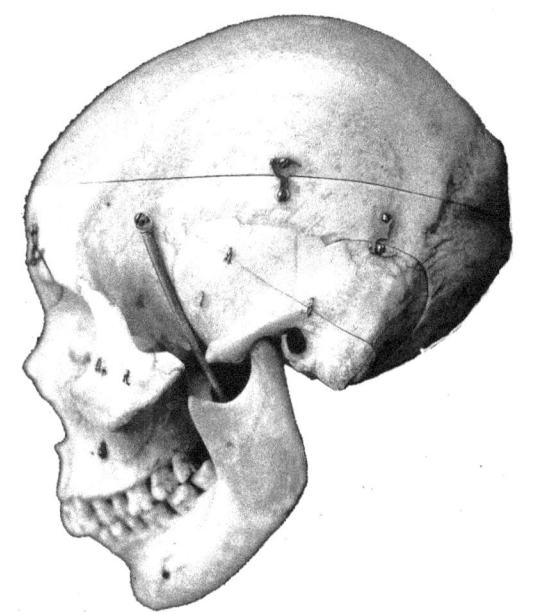

When students were pared off to model portrait heads of each other, I found they had difficulty understanding the head's structure. Their preconceived notions were getting in the way.

I borrowed a human laboratory skull from the biology department at Lycoming College where I taught and took it home for the summer. I wanted learn how to show my students that the head was a functional object, to offer them an escape from their preconceived imagery. I studied the skull using the Bauhaus dictum that "form follows function."

The Bauhaus, a German art school that operated between the world wars, was very influential on modern design. The emphasis was in having the design of an object or building reflect its function. This was also the logic that Andreas Feininger explored in his photo essay book <u>The Anatomy of Nature</u> which I greatly admired and used in my teaching. Andreas Feininger, I later learned, had been a student at the Bauhaus from 1919, a time near is its inception, and remained associated with them for years. He later came to the U.S. and worked as a photographer for Life magazine.

In The Anatomy of Nature, Feininger took the Bauhaus ideals and applied them to nature showing the convergence of form in diverse natural structures due to similar forces and functions. Steeped in The Anatomy of Nature, I expected to be able to read the functions of the human head from the skull sitting on my drawing table. This turned out to be both true and a very radical approach which has put me at odds with the literature of paleontology, physical anthropology, and paleoanthropology.

I often asked my students to touch their own faces to feel and understand the forms they were trying to see, but they looked at the clay already on their hands and usually decided against it. This was a lost opportunity because our sense of touch and our sense of sight are closely related. Visual knowledge can be built directly from our sense of touch.

I taught my students to see that the human brain case is egg shaped. Unless I made a point of it, they would ignore the brain part of the head and focused primarily on the face with predictably bad and distorted results. Functionally, structurally, and aesthetically, the head makes no sense without its cargo of brains. In being egg-shaped it is like a number of other shell containers in nature where a soft, fragile substance needs protection from the hard knocks of the outside world.

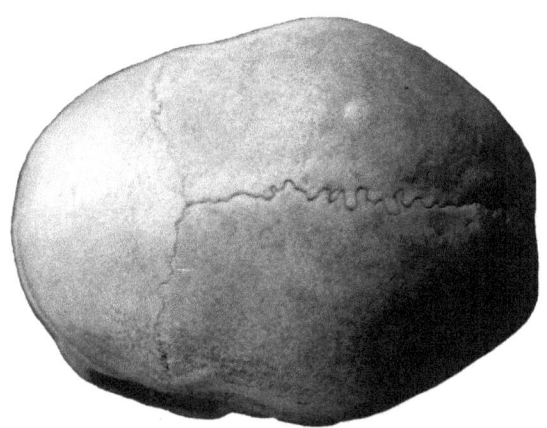

You can experience this in your own brain case by placing the heels of your hands on your temples and stretching your fingers back along the side of the skull. You will feel the volume expand as your fingers reach rearward. Our foreheads are the small butt end of the egg, while the larger end is the back of the head. All communication reaches the brain through ports that penetrate the underside of the egg. These includes the pathways for sensory nerves from the eyes, ears, and nose; the foramen magnum, the big hole for the spinal cord, and numerous other passages for nerves and blood vessels.

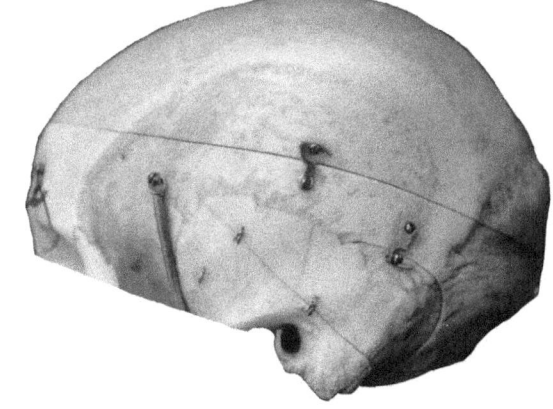

The underside of the egg shaped brain case can be found in a head or skull seen in profile by a line connecting the bridge of the nose, the corner of the eye, the opening of the ear canal, and extending to where the back of the neck and skull meet. Anatomists will find this imprecise but it is a good approximation and can be used to easily locate the floor of the brain case in living subjects

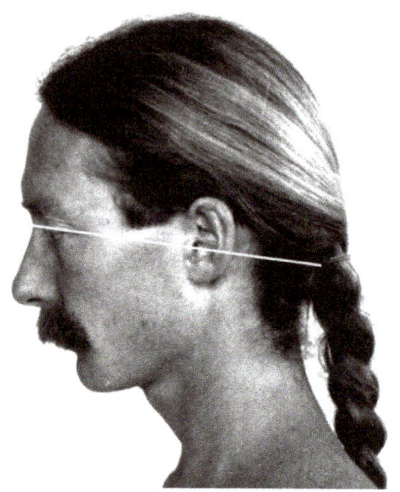 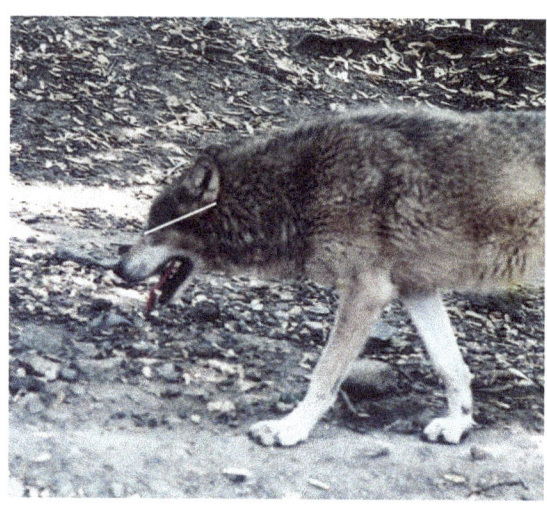

The two lines denoting the base of the brain case show a distinction between bipedal and quadruple animals. In the human head the line descends from bridge of nose to ear while in the wolf it ascends front to rear.

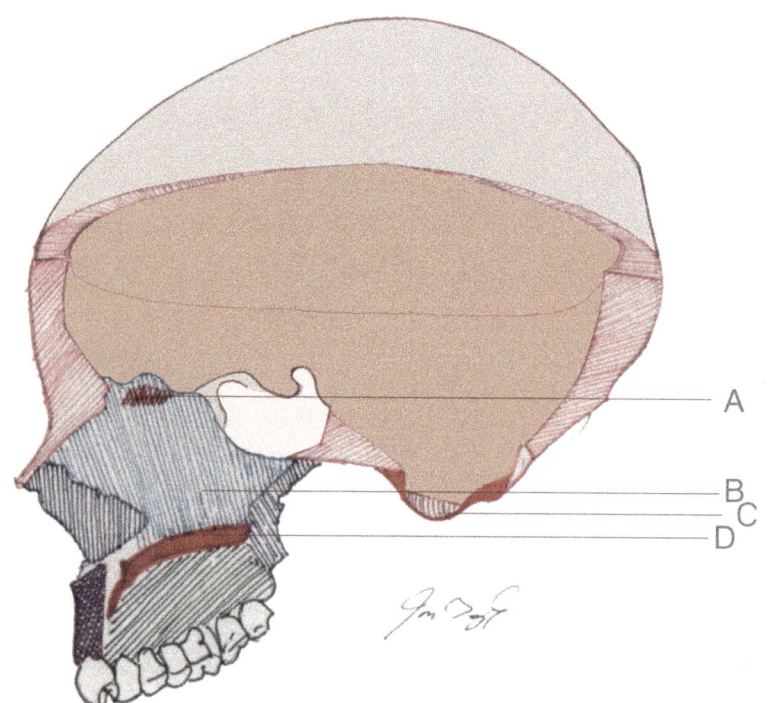

Drawing of the laboratory skull split along the midpoint showing the elements of the braincase and the facial stalk. A. The cristas gallus, with the olfactory nerves extending through the base of the braincase. B. The septum of the nose. C. The foramen magnum, the entrance for the spinal column in the base of the braincase. D. The hard palate.

passageway arches through the facial stalk from front to rear, carrying air to the under side of the brain case where the olfactory nerves monitoring for smell are located.

The facial stalk contains the eyes, the entire apparatus for the nose, upper jaw and hard palate, and encompasses the support structures used to handle the forces created in chewing and wrenching with the teeth. The facial stalk carries all the primary senses except for hearing and balance.

Where the facial stalk flows out of the brain case, a single sheet of bone forms the upper portion of the nose and the inside surfaces of the eye socket. The facial stalk's lower end is the upper dental arch, the immovable upper jaw. The hard palate, a bone plate, fills the upper dental arch where it helps the tongue direct food and keeps food from being pushed upwards into the soft nasal cavity during chewing.

Human eyes are larger than we think, being about the size of a ping pong ball. The eyes are deeply set under the front portion of the brain case protected by the surrounding bones of the eyebrow, nose, and cheek. Eyelids cover much of the eyes adding to the illusion that the eyes are smaller than their actual size. The interior top surface of the eye socket shares a common wall with the floor of the brain case. To its inside, as mentioned above, the eye socket also has a common wall with the nasal cavity. The lower surface of the eye socket is formed by one of the four sides of the zygomatic bone which contains the lower sinuses.

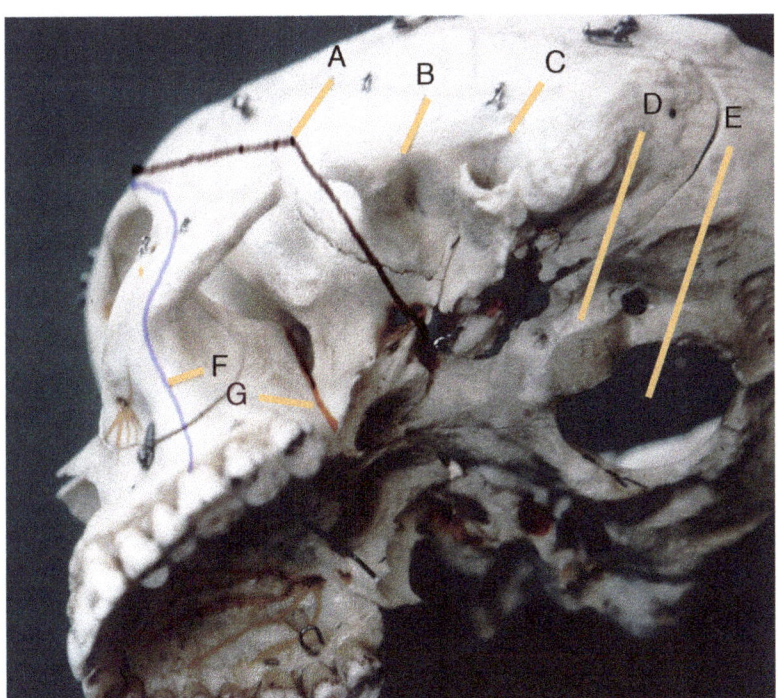

A---- *Red line approximates the boundaries of the facial stalk.*

B---- *Lower jaw joint.*

C---- *Ear Canal*

D----*Condyle that forms the joint with the first vertebra.*

E---- *Foramen Magnum, the entrance for the spine.*

F---- *Blue line traces path for molar stress.*

G----*Orange Line marks the edge of the pillar that acts as an anvil for the upper teeth.*

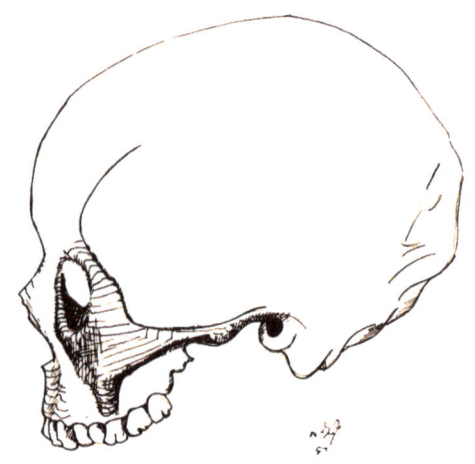

Zygomatic complex including the post optical bar which carries molar pressure.

The zygomatic bone is a tetrahedron, a four-sided pyramid whose hollow interior forms the large sinus below the eye. The base of this pyramid has a common wall with the nasal cavity. One of the other three remaining sides form the lower inside surface of the eye socket as mentioned above. You can feel one of the two remaining sides if you place your finger along your nose and move it horizontally towards the outside of your head. If you poke around a bit while you do this, you will discover that this bone gets smaller as it moves away from the nose. Just when you are expecting it to come to a point, it turns and moves rearward towards the ear.

Continue tracing its progress with your finger. As it moves rearward you can feel it end just above the ear tab in front of the opening to the ear canal. This section is a flying buttress form. Starting at the cheek bone it arches outward away from the head and reattaches to the skull near the ear canal.

In the human head, the zygomatic arch has the same function that it serves in other vertebrates: It transfers to the skull the lateral stresses generated by the teeth of the upper dental arch when we wrench and tear sideways at food or other objects.

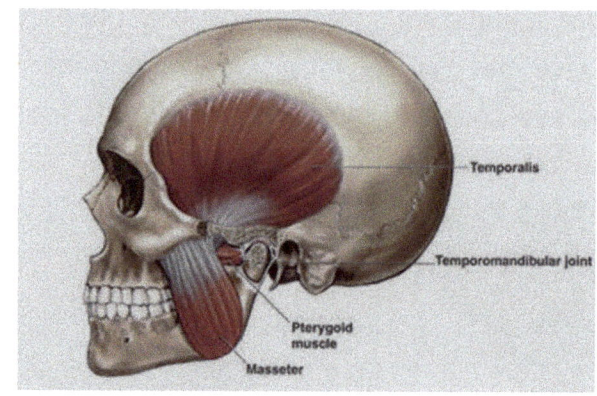

Proprofsflashcards.com

The temporalis muscle is one of two primary muscle groups that pull the jaw closed. The temporalis attaches to a wing-like bone projection on the jaw, the coronoid process, and then fits under the arch form of the zygomatic complex to anchor to the temple area of the skull. If you clench your teeth while you touch your temples and you can feel the temporalis muscle contract. The masseter muscle also pulls the jaw closed by wrapping around the outside of the jaw near the ear and anchoring to the lower edge of the zygomatic complex. You can feel the masseter if you push in just below your cheek bone while you clench your teeth.

The jaw hinges into a depression just to the front of the openings of the ears. This depression, called the mandibular fossa, is an integral part of the zygomatic arch where it attaches to the skull. For our purposes, we can think of the jaw as having a hinge joint. While this is primarily true, the jaw has a remarkable amount of articulation. If you open your mouth slightly, you will find that you can move your lower jaw sideways while thrusting it in and out. These movements, which we use to align our jaw when chewing and grinding food, are made possible by the shape of the cartilage that forms the joint and to a variety of small muscles inside of the jaw frame.

Just to the rear of the cheek bone, where the zygomatic turns to move rearward, the bone forks, with one spur growing upward to connect to the eyebrow ridge at the front of the temple. This spur is the post optical bar which forms the outside margin of the eye socket. From this edge, a cup shaped bone flange grows inward to support the eye. It is easy to feel the post-optical bar as it diverges from the zygomatic arch at the corner of the eye. If you press in just behind the post-optical bar at the temple, you can often feel the eye socket from behind.

All primates, not just humans, have their eye sockets (optic orbits) completely ringed with bone. This is part of the scientific description of a primate. Dogs, however, do not have this feature. Their eyes are against the side of their brain case and their eye sockets are open to the rear. Nevertheless we can experience a heartfelt eye to eye communication with our dogs because they, too, have frontal eyes.

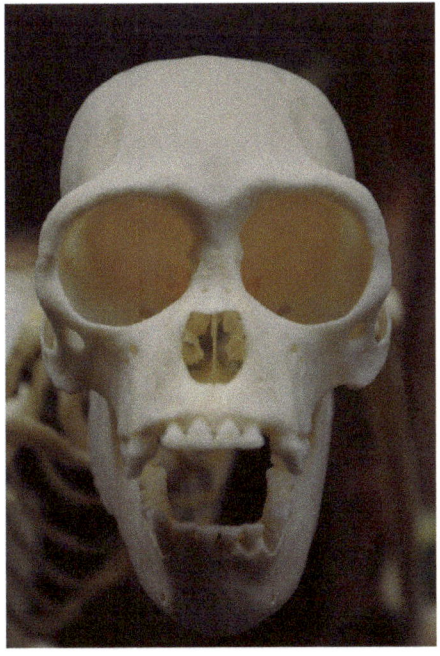

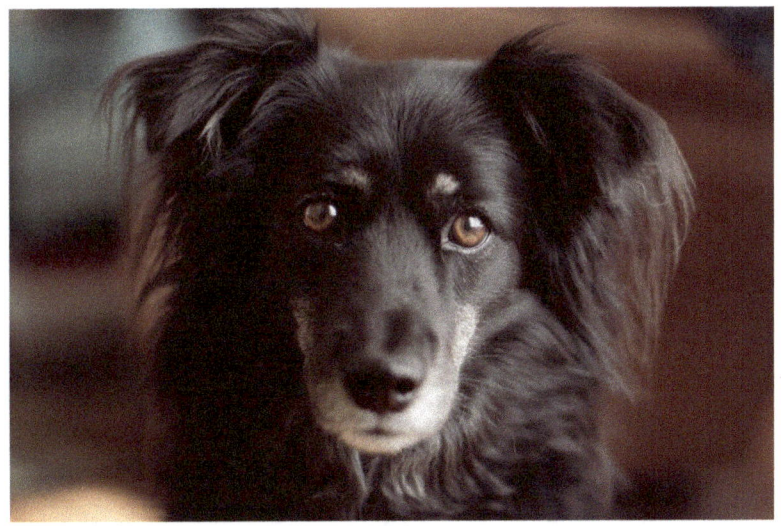

Spider monkey skull, Courtesy: Museum of Osteology, Oklahoma City
Dog Photo Chung-Nguyen Courtesy Unsplash

The floor of the brain case continues forward over the facial stalk. In an articulated lab skull, it is possible to remove the cap of the skull and see the area that is above the eye sockets. Between the eyes there is a spongy area—the cribriform plate, which is split in the middle by a thin raised flange of bone, the crista galli (L. cock's comb). The crista galli is the extension of the septum of the nose continuing into the interior of the brain case. It separates the two spongy areas through which the olfactory nerves penetrate. Separating the two airflows from the nostrils producing a directional indication for the origin of smells.

To the rear, the floor of the brain case contains the foramen magnum, the entrance port for the spinal column. To either side of the forward portion of the foramen magnum are two bone lobes, the occipital condyles. The first vertebrae below the brain case has depressions that ride in these condyles. This first vertebrae is called the atlas vertebrae because it holds up the head referencing the classical god Atlas holding up the world. The occipital condyles and the atlas vertebrae form the joint that we use when we nod our heads. The brain case floor also contains the structures for the ear canals, which in turn hold the organs of balance.

This forward part of the floor of the brain case is also the roof of the facial stalk. It contains

Crista Galli
    Is an extension of the septum of the nose that surfaces inside the brain case floor and separates the two cribriform plates.

Cribriform Plate
    Is a spongy area where the olfactory nerves penetrate.

Anterior Cranial Fossa
    It carries the frontal lobes and is the roof of the facial stalk. Underneath, it forms part of the eye socket.

Optical Canal
    Where the optic nerve enters the brain.

Auditory Canal
    Port for the auditory nerve. The working of the inner ear are imbedded in the floor of the brain case.

Foramen Magnum
    Where the spine enters the brain case.

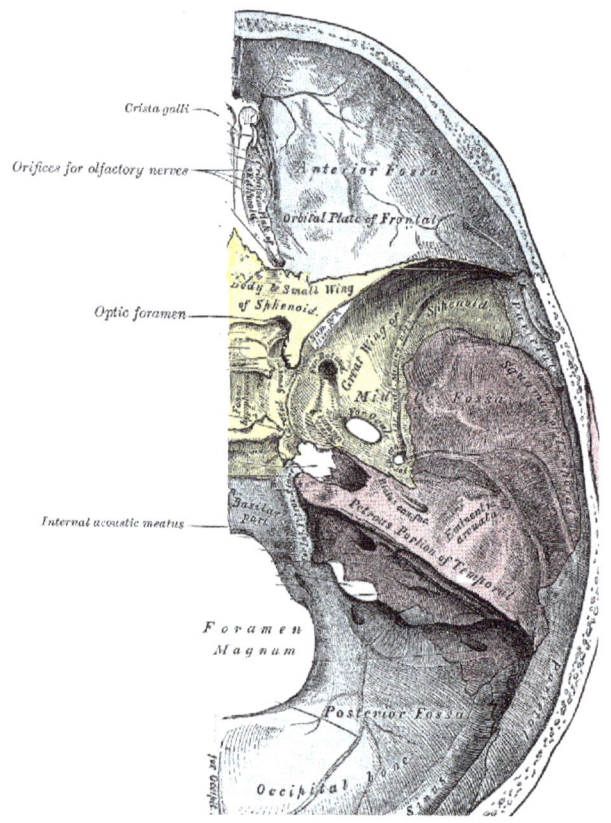

Henry Vandyke-Henry Carter 1858 Wikimedia

the upper potion of the eye sockets and the olfactory nerves of the nose. When the facial stalk rotates inward or outward, this portion of the brain case floor rotates with it. In humans, the forward portion is angled in relationship to the rear floor of the brain case but it is less so in other great apes.

A great deal of scholarly study has been invested in determining why. One group thinks it is the size of the face relative to size of the brain that is the cause. Another suggests that enlarging brains, particularly the enlarging temporal lobes, have caused the face to rotate under the brain case. This part of the brain case floor, however, is also the top of the facial stalk and it follows the stalk when it swings inward or outward. In humans our vertical posture has caused the facial stalk to swing under the braincase. The front portion of the brain case, the roof of the facial stalk has a large angle break with the rear of the brain case floor. humans, the facial stalk swings inward or outward to set the face properly into the visual environment. The overlay of the Inuit and Congolese skulls on page 54 demonstrates this. The facial stalk's reset is achieved by bone accumulation or absorption in the frontal portion of the skull. Following is an excerpt from a summary of an article by a set of prominent authors.

MIDDLE CRANIAL FOSSA

Middle Cranial Fossa Anatomy and the Origin of Modern Humans

MARKUS BASTIR,
ANTONIO ROSAS,DANIEL E. LIEBERMAN, AND
PAUL O'HIGGINS

"Anatomically, modern humans differ from archaic forms in possessing a globular neurocranium and a retracted face and in cognitive functions, many of which are associated with the temporal lobes. The middle cranial fossa (MCF) interacts during growth and development with the temporal lobes, the midface, and the mandible. It has been proposed that evolutionary transformations of the MCF (perhaps from modification of the temporal lobes) can have substantial influences on craniofacial morphology....

In particular, the findings of this study point to variations in the temporal lobe, which, through effects on the MCF and face, are central to the evolution of modern human facial form."

2008 Wiley-Liss, Inc. Harvard University DASH

So here is the problem: These authors follow a long scholarly tradition of tracking traits in trying to define which are uniquely human. The angled base of the human brain case fits the bill and therefore must point to an advancement towards being human. In this case the temporal lobes of the brain have disproportionately enlarged. According to the article it was our enlarging temporal brain lobes that caused this rotation. This is judged by the size of the temporal lobe's depression in the brain case floor which is referred to as the middle cranial fossa or MCF.

The temporal lobes enlargement, they speculate, rotated the face rearward and caused our uniquely angled brain case floor. In reality, however, it is our uniquely bipedal stance and posture that created the change. Missing is the recognition that the face has a static relationship to the sensory environment. The face is fixed in space while every other aspect of the skull changes to keep that relationship. As mentioned earlier, prognathism is an unfortunate illusion of the Table Image Fallacy. Aiello and Dean show an illustration on page 196, figure 11.1 of Huxley's 1863 text showing reduced prognathism in humans versus ape, lemur and beaver. On the next page, figure 11.2 is a drawing of the Gorilla Growth Series placed in the Frankfurt Horizontal position to demonstrate enlarging prognathism as the gorilla's mature. Faulty illustrations have distorted both the imagery and the research.

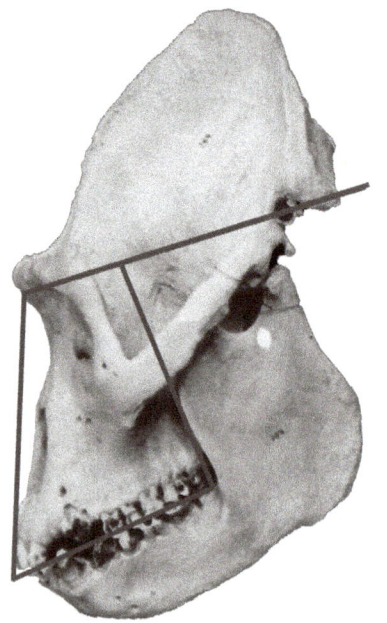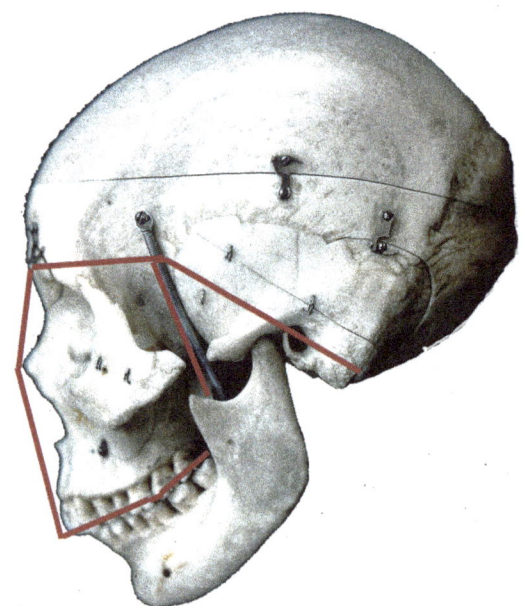

In both the adult gorilla skull and the human lab skull, lines indicate the floor of the brain case. In the gorilla it is a continuous plane but the human has a break from front to back. In both cases the face stays fixed in space but the human brain case has rotated downward following the changing angle of the more vertical spine. The angle of the floor of the human brain case doesn't reflect our large brain but rather our bipedal stance.

MOUTH

The mouth is a hammer and anvil with the upper teeth being the anvil and the lower jaw the hammer. Mammals use the bone shell of the nasal cavity to absorb the forces on the upper jaw. Primates, however, have two separate buttresses for these pressures. The front of the mouth uses the bone shell of the nasal cavity like other mammals. The molar region forces, however, are carried through the zygomatic complex which is centered above the first molar and then along the outside edge of the eye socket into the brain case. When the skull and jaw are aligned in a way that these forces are projected in front of the skull, then eyebrow ridges enlarge to absorb these pressures.

This configuration is a compromise that gives us our frontal eyes but limits our peripheral vision to the rear. The literature is full of assertions that frontal eyes are an adaption for judging distances in our tree dwelling, branch leaping past. Is that the case? Maybe we should ask a squirrel flinging itself through the air to land on a branch or a sparrow flying through tangled branches and then landing expertly on a wire.

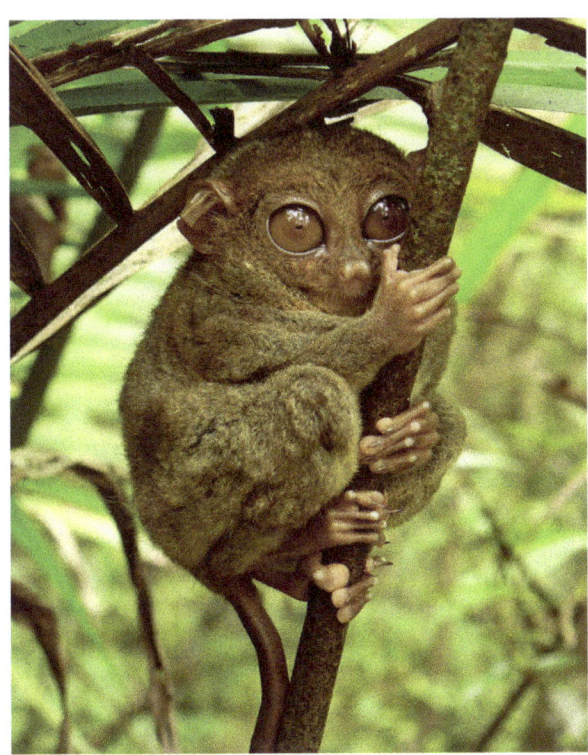

We inherited this problem from a distant tarsier-like ancestor. Animals that have their eyes ringed with bone are nocturnal sight hunters like the huge eyed owls and tarsiers. Owls and tarsiers are a convergent form. Both

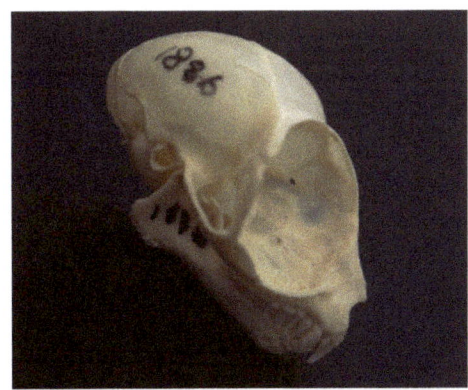

have their huge eyes ringed with a supportive bone flange and both have eyes that are fixed in their skulls and, to compensate, each has a neck that can rotate their head almost 360 degrees. The tarsier's eyes are so large that the shells of the orbits meet in the middle which disrupts the flow of mastication pressures through the nasal bones.

The eye flange has completely suppressed the zygomatic complex and pushed it down to the margin of the upper molars. The tarsier shanghaied its eye flange to carry dental forces and thereby rewrote the engineering specs for later primates.

House cats are also nocturnal hunters with large eyes and they, too, have bone flanges that almost encircle their eyes. Unlike primates, however, the bones growing from the temple region and up from the zygomatic complex don't quite meet because cats aren't using their eye flanges to conduct mastication pressures.

Animals that operate in the daytime don't need supporting bone rings around their normal sized eyes. Nature's rule is to repress unused structures to save energy. All primates have a post optical bar because it has a function. In the last few years an almost complete fifty-five million year old fossil of a tarsier-like creature was found in China. This fossil was quite like living tarsiers except for its lower hind legs which didn't have the elongated tarsus bone that gives tarsiers their name. This fossil predates all monkeys and apes so is a possible antecedent of primates.

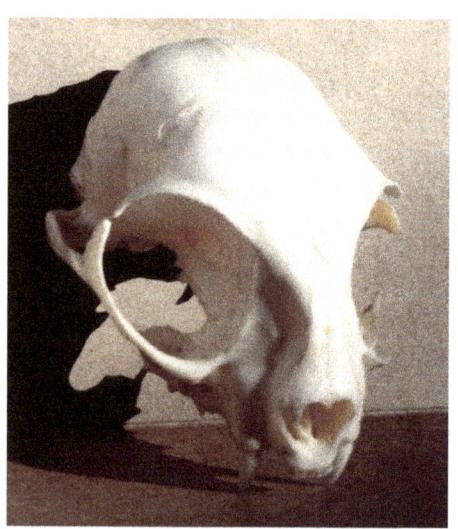

Cat skull, the post optical bar isn't continuous. The cat being small and nocturnal has large eyes but doesn't use the supporting optical bar to carry dental forces.

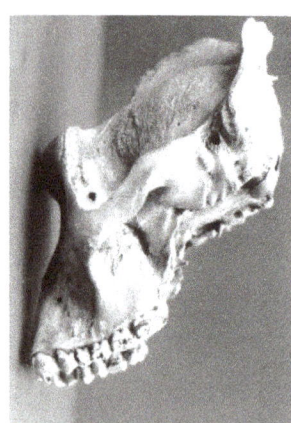
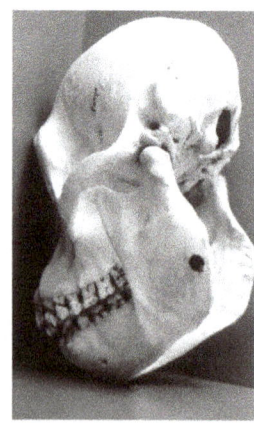

Gorilla and orangutan photographed with strong light showing the bone line from above the first molar to the eyebrow ridge. The human lab skull shows the same mechanical structure but it terminates in the shell of the brain case, not an eyebrow ridge.

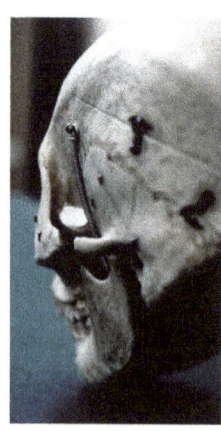

JAWS

Jaws are amazingly misunderstood. The literature is full of assertions that modern humans' small jaws and teeth are the result of eating soft food. Jaws are third class levers. A first class lever has the weight to be moved at one end close to the fulcrum while at the far end is the applied force. I still have my family's pry bar. I don't know how old it is or why I still have it but with just a brick for a fulcrum it will budge quite a large rock.

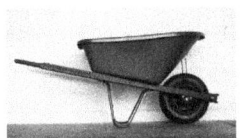 A second class lever has the fulcrum at one end, then the load, and at the other end the applied force. A wheelbarrow is a second class lever. The handles move farther than the load which gives a mechanical advantage. Lifting the handles requires considerably less force than the weight in the barrow.

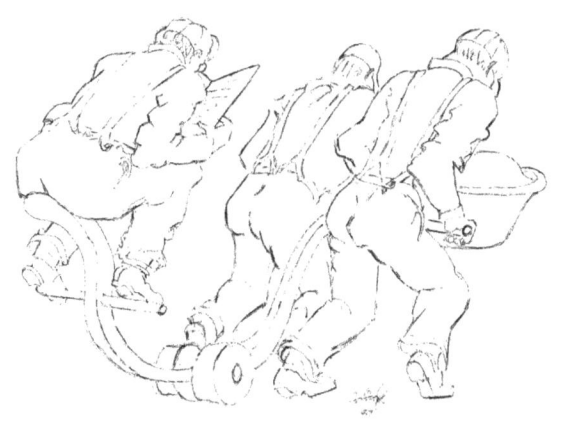

In this wheelbarrow the muscle power is between the load and the fulcrum as it is in the jaw.

A third class lever has the fulcrum at one end near the force with the load at the other end. Tennis rackets, baseball bats, and fishing rods are third class levers. They don't multiply force but instead they multiply speed. Imagine having a bag filled with fifty pounds of stuff. Now take a baseball bat and put the fat end through the bag's handles and try to pick it up. Holding the bat in the normal batter's position would make the task virtually impossible. Sliding one hand to the middle of the bat helps the situation. It will still require much more that fifty pounds to raise the bag because the load is travelling further than the lifting hands.

That is roughly the situation with jaws. They are structured to be like a fishing rod amplified for speed not power. Perhaps some ancient ancestor found it advantageous to be able to quickly snap its jaws shut and we inherited its mechanics. For us and other mammals the jaw is inefficient for producing biting power especially at the front of the mouth. The short jaws of modern people are more efficient than the longer jaws of apes and some ancestors.

Humans have small molars which, again, is widely reported in the literature as evidence of eating softer food. At one point in my sculptor career I had the idea that I could stamp out my designs in a press. A friend let me use his twenty ton hydraulic press and I fabricated the dies to use in forming sheets of aluminum. What I quickly learned was that the larger the work the less deeply the press would imprint. Larger surfaces lessen the pressure while small surfaces concentrate it. Goldsmiths can weld gold sheets together with just their hand pressure using a tool with a small flat end.

Due to their longer jaws, our ape relatives need large powerful muscles to obtain the needed biting force at their front teeth. Male gorillas grow a crest along the midline of their skulls to give more area to attach their temporalis muscles. The molar region of the mouth is where the major jaw muscles are located. Molars being aligned with the muscles are more mechanically advantaged than is the front of the mouth so the chewing pressures there are considerably greater. Teeth have a finite material strength and the gorilla's large molars spread the force out and save their molars from being shattered. When all this is considered, the bite pressures between human and gorilla molars is likely quite similar.

Whether jaws are longer or short is a factor of posture not mastication. Animals with horizontally held spines have long jaws while ones with vertically held spines (humans) have short jaws. More on that later.

GORILLA GROWTH SERIES

The gorilla growth series, a photograph by Uno Samuelson of the Stockholm Museum of Natural History, is wonderfully informative and has been used a number of times in the literature of evolutionary anthropology. It shows a series of five gorilla skulls in sequence as they mature from infant to adult male. The skulls, following standard practice, are placed in the Table Fallacy position with the jaw resting on the shelf and the brain case propped to support them in this position.

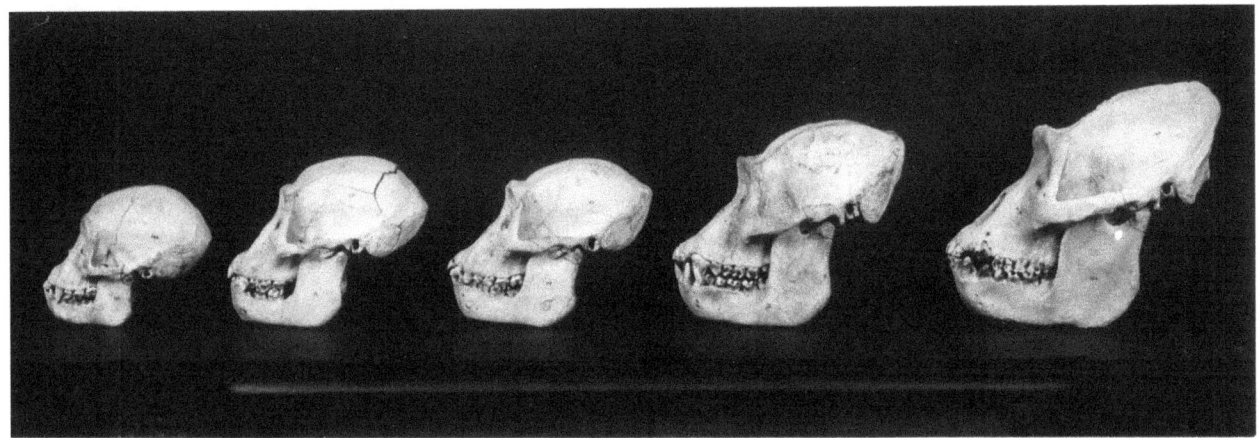

Photo: Uno Samuelson, Stockholm Museum of Natural History

Samuelson dramatically put the skulls in a black environment so they seem to float in the picture. We can discover a number of things about the maturing process in gorilla heads by studying Samuelson's photograph. The face grows larger and is also larger in proportion to the brain case as gorillas mature. An eyebrow ridge, which is absent in the most juvenile head, appears in the second head and grows progressively more pronounced as the heads in the series mature. The nuchal (or occipital crest) in the back of the skull becomes more pronounced and flatter its angle raises from fairly horizontal in the juvenile to more vertical as maturation progresses. The temple line, which denotes the margins of the temporalis muscle, progressively creeps upward on the brain case and finally occupies the sagittal crest on the most senior head.

What I've noted are the changing attributes of gorilla heads as they mature. What we don't know from Samuelson's photo is why the gorilla radically changes its head structure as it grows. To find that out we have to break the hold of gravity on these skulls and move them out of the Table Fallacy position. I have used Photoshop to reorient the skulls so we can see them in the same attitude that gorillas would have held their heads during life. This new positioning reveals a set of design imperatives driving the structural changes that occur during maturation.

Finding the right position for these skulls only requires applying two well worn rules: the artist's rule for placing eyes half way from the top of the head's image to the chin, and the genetic program's rule that the nose always leads. Although I feel rather confident about this realignment,these skulls were from five separate gorillas and not growth stages of a single individual. Gorillas, like us, are individuals.

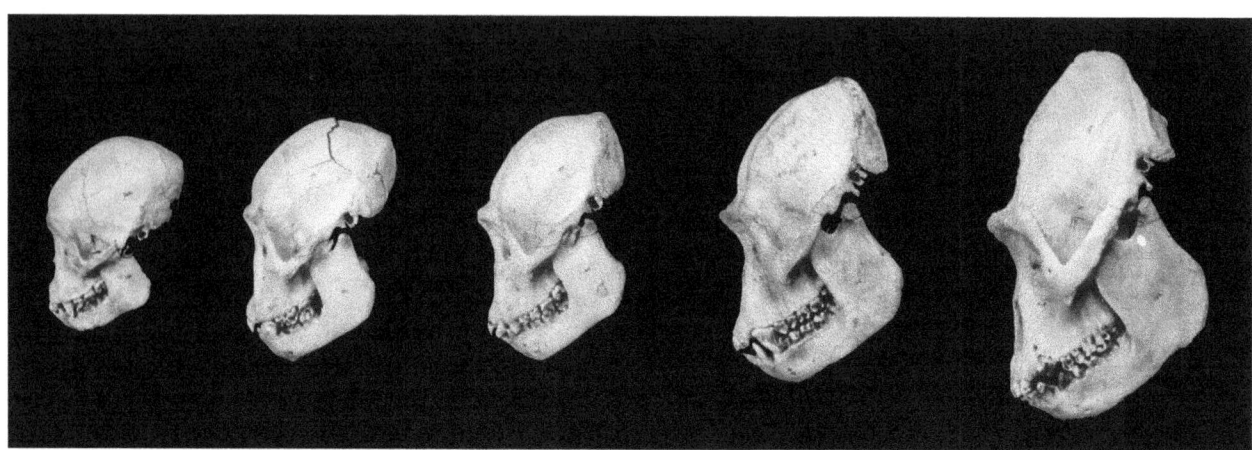

Samuelson's skulls have been rotated to the position they were held in life as As the gorilla moved forward alertly, the relationship of the face to the environment is fixed. Other design elements of the head are modified to preserve this relationship as the gorillas mature from the infant's near vertical bipedal posture into a horizontal quadrupedal orientation.

I can locate the center of the eyes in a skull by finding the bridge of the nose. Animals hold their heads so they can sight across the low point of the nose to get the largest field of vision possible. You can check this for yourself by holding your head in the normal position you would use when walking or sitting and then, closing one eye, and look across your nose with the other. You will find that your line of sight is level with the notch of the bridge of the nose.

Another notch designed to increase your field of vision is where the post optical bar meets the cheekbone. If you put the tip of your index finger at the outside corner of your eye, you will discover that it blocks your peripheral view to the rear and downward. This design detail helps to keep them from sneaking up from behind. Now, if you run your finger on the bones surrounding your eye, it will outline your optical field.

In the drawing on the following page, the gorilla's face is quite similar to the model's, but the gorilla's ear, being an element of the brain case and not of the face, has risen and rotated with the changing relationship of face to brain case. Furthermore, both the optic bar on the

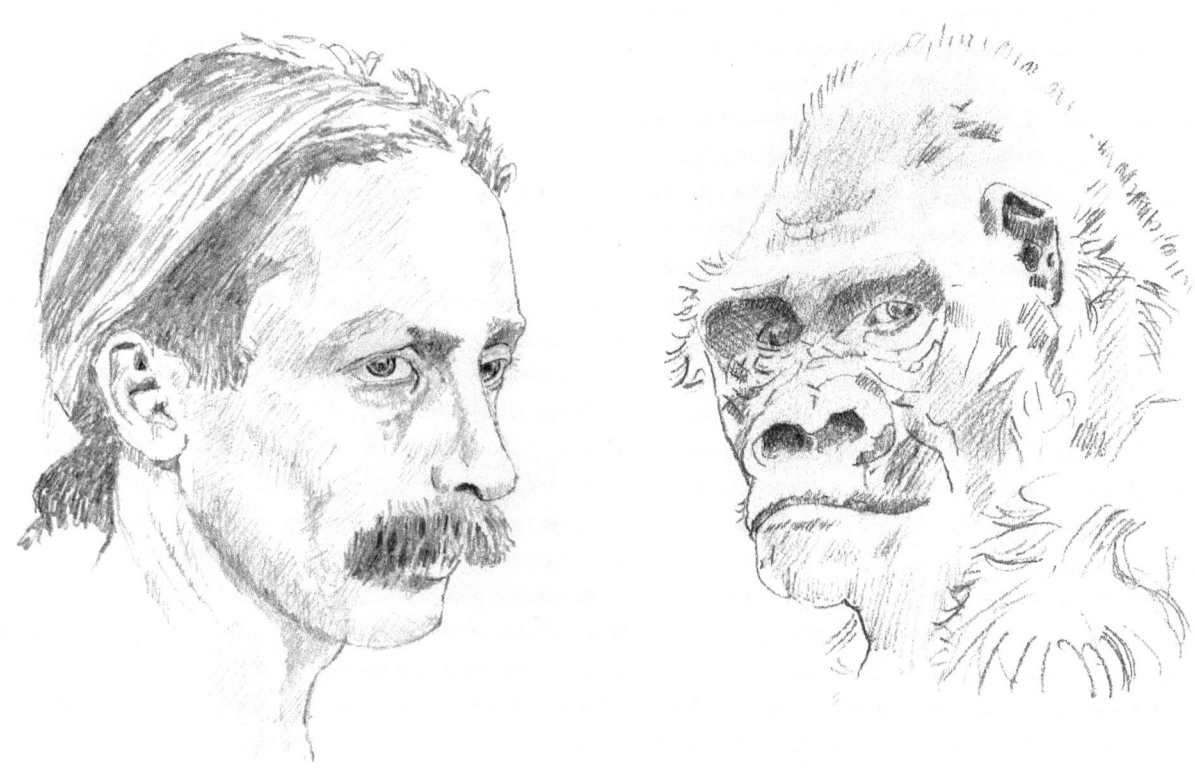

In the drawing above, both the gorilla and the model sight across the bridge of the nose and face forward alike.
The gorilla's ear, however, is higher and rotated because ears are located in the base of the braincase and move
with its changing relationship to the facial stalk.

outside edge of the eye socket and the cheek bone (the zygomatic complex) form part of
the eye socket, but these are also primarily mechanical struts used to carry dental pressures. This dental support function, however, keeps the zygometic complex from performing its unlimited expression because the eyes must get as large a visual field as possible.
The final design is determined by these vital functions interfering with each other.

Composite photo demonstrating the restructuring of gorilla heads during maturation.

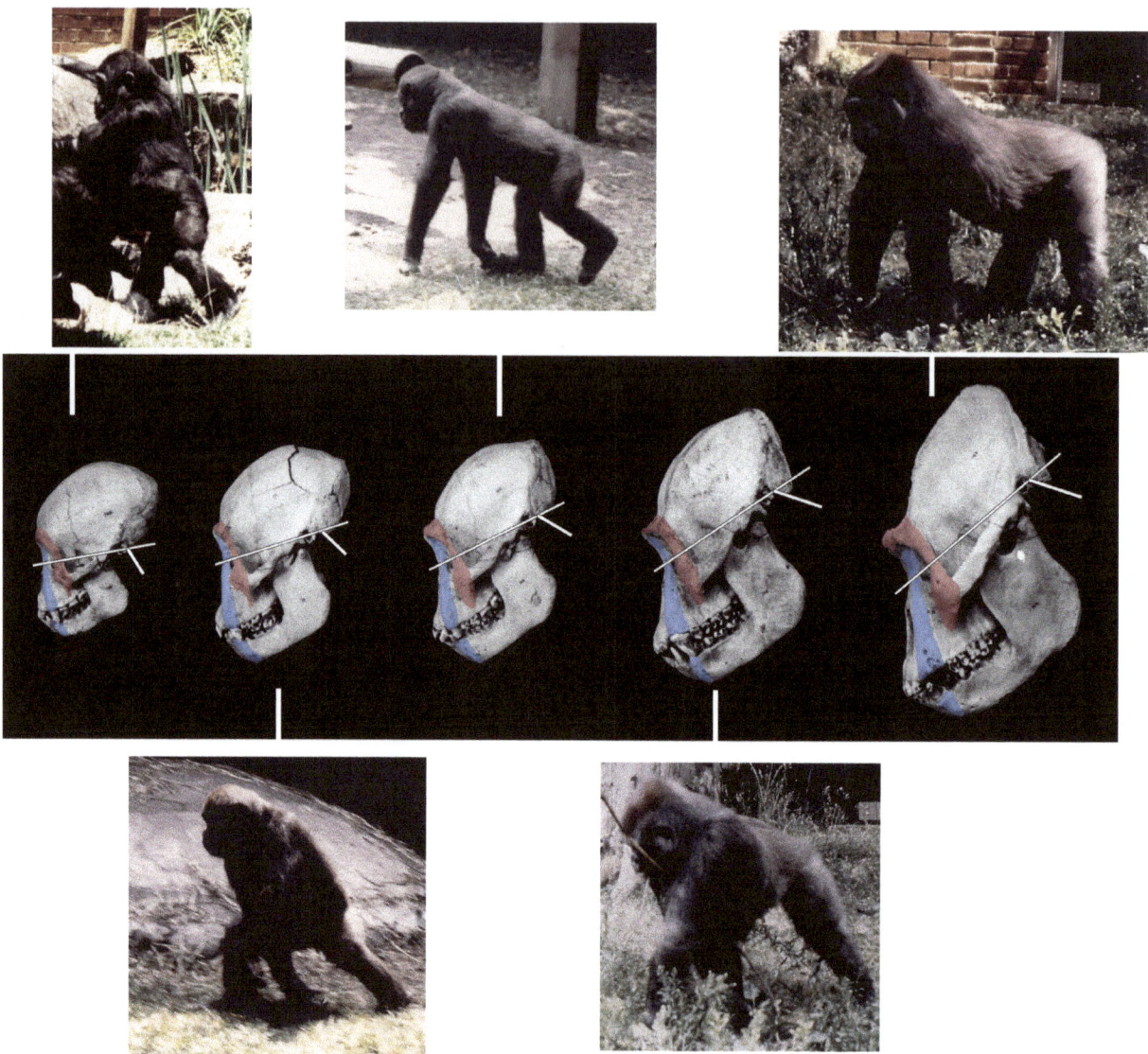

Two lines have been added to the modified Gorilla Growth Series: one denotes the underside of the brain case, which can be found by connecting the bridge of the nose and the ear canal. The other indicates the angle of the spinal column as it meets the base of the skull. This angle, which is obtuse to the front and acute to the rear, is fairly constant throughout the growth cycle and is approximately the same angle that our human spines have to the base of our brain cases. For each skull there is a picture of a living gorilla of the same approximate age. These photos show that the gorilla's spine is carried more horizontally as the animal matures. Notice that at all ages the gorillas hold their faces in the same position as they go forward.

The two youngest gorillas are walking in a more or less bipedal mode, the youngest holding on to its mother. Like our own young, they spend a good deal of time being held upright by their mothers, and their faces are configured for this upright position. Our children stand up and begin to walk on two legs as soon as they are able, because they can't see where they are going if they are down on all fours. Gorilla babies have the same problem, so they also get up and move about on two legs. The difference is that we stay vertical as we mature but gorillas become increasingly horizontal and their heads configured so they can be on all fours and still see where they are going.

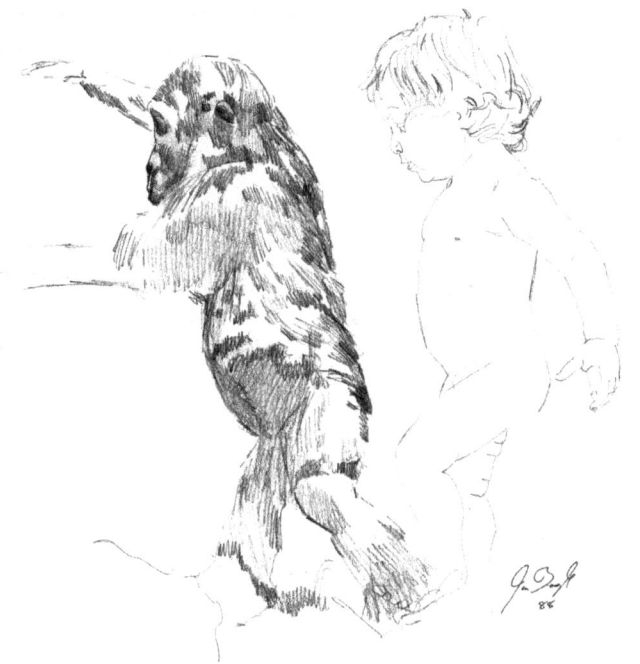

These two youngsters are walking upright, the baby gorilla holding on to its mother so he can see. Human babies will sit up before they can stand or stand clinging to an object, like the gorilla baby, also to see and to relate to the space around them more comfortably.

A number of things are happening in the photos of the gorillas. Notice that the spine goes from being almost vertical in the youngest to being almost horizontal in the senior adult. As the spine rotates downward, the brain case moves with it as a unit, like a lollipop on a stick.

The face, however, is fixed in space by its functional relationship to the external environment and can't rotate with the brain case. It is locked in space as the brain case rotates upward the angle between the floor of the brain case and the facial stalk opens. The jaw hinges into the base of the brain case and is also locked into a relationship with the front of the face. It stretches out, becoming longer and more mechanically inefficient in the process.

Force is needed at the incisors to provide cutting power. The muscles of the jaw, however, are attached nearer the hinge joint, so this incisor force becomes increasingly harder to deliver as the jaw grows longer. With the jaw increasing in size, its attendant muscles, along with their points of attachment, also grow to keep in proportion. We sculptors know--from sometimes painful experience--that when the size doubles, the surface area quadruples and the volume of a mass is cubed. An simple increase in size has exponential results.

The face continues to grow while the brain case remains fairly constant in size. In the Gorilla Growth Series we can see the line of attachment for the temporalis muscle creep upward on the skulls until, in the most senior head, it requires an added sagittal crest to anchor it.

The red area in the skulls denotes the flow of the dental forces coming off the molar region of the upper jaw. The pressure from the region of the first molar is conveyed up the zygomatic process and is carried by the post optical bar on the outside of the eye. In the juvenile skull these forces are anchored and dissipated in the bone shell of the brain case but as gorillas mature, the forces go to a point in front of the brain case. The increasing size of the eyebrow ridges in gorillas is compensation for the increasing misalignment of these forces while the jaw is producing even greater pressures in the molar region.

The blue area on the skulls denotes the dental pressure from the front of the mouth. Animals use the bone shell of the nasal cavity to carry and dissipate dental pressures, but in primates, just the forces from the front of the mouth use this pathway. In primates the buttress form above the canine teeth can be seen rising past the margin of the nasal opening, while in the lower jaw, the buttress forms descend to meet at the chin.

ORIENTING HEADS

If we are going to use the living animal in a systematic comparison of skulls we need to know how living animals hold their heads.   Animals orient their heads precisely in the environment to overcome the slow workings of a chemical nervous system. There are three significant positions that determine the design of skulls: the going-alertly-forward position, the high-head-surveillance position, and the head-down-feeding position. These are precise, they relate directly to the survival of the animal and they dictate the design of many of the bone lines of the skull. Of course, an animal can move its head in a myriad of ways but most of these positions are transient and not part of a consistent survival function.

The head-down-feeding activity obscures environmental smells but increases peripheral vision to the rear. You can try this yourself. Hold you head normally and make yourself aware of the range of your peripheral vision now lower your face so it is roughly parallel with the floor. Like a deer feeding, you will find your peripheral vision to the rear is greatly expanded.

This is partly due to a design feature of our heads and of other animals. I call this feature the "Oral Optical Groove." Place your finger tip on your lips and slide it out to the corner of your mouth. You find that you will lose the peripheral view of it just beyond the corner of the mouth. The Oral Optical Groove allows for animals to visually monitor their feeding activity. Artists teach their students to have the corners of the mouth be directly under the pupils of the eyes which helps establish the proportions of the mouth.

Animals, in the high head surveillance position, stand still and raise their heads to pick up sensory clues at a distance. Standing still removes their own sound and movement allowing them to be able to better locate the source of the event. Grazing animals take turns using all three head positions to get maximum sensory coverage for the herd. We humans also stand still and raise our heads when confronted with a surprising sensory event.

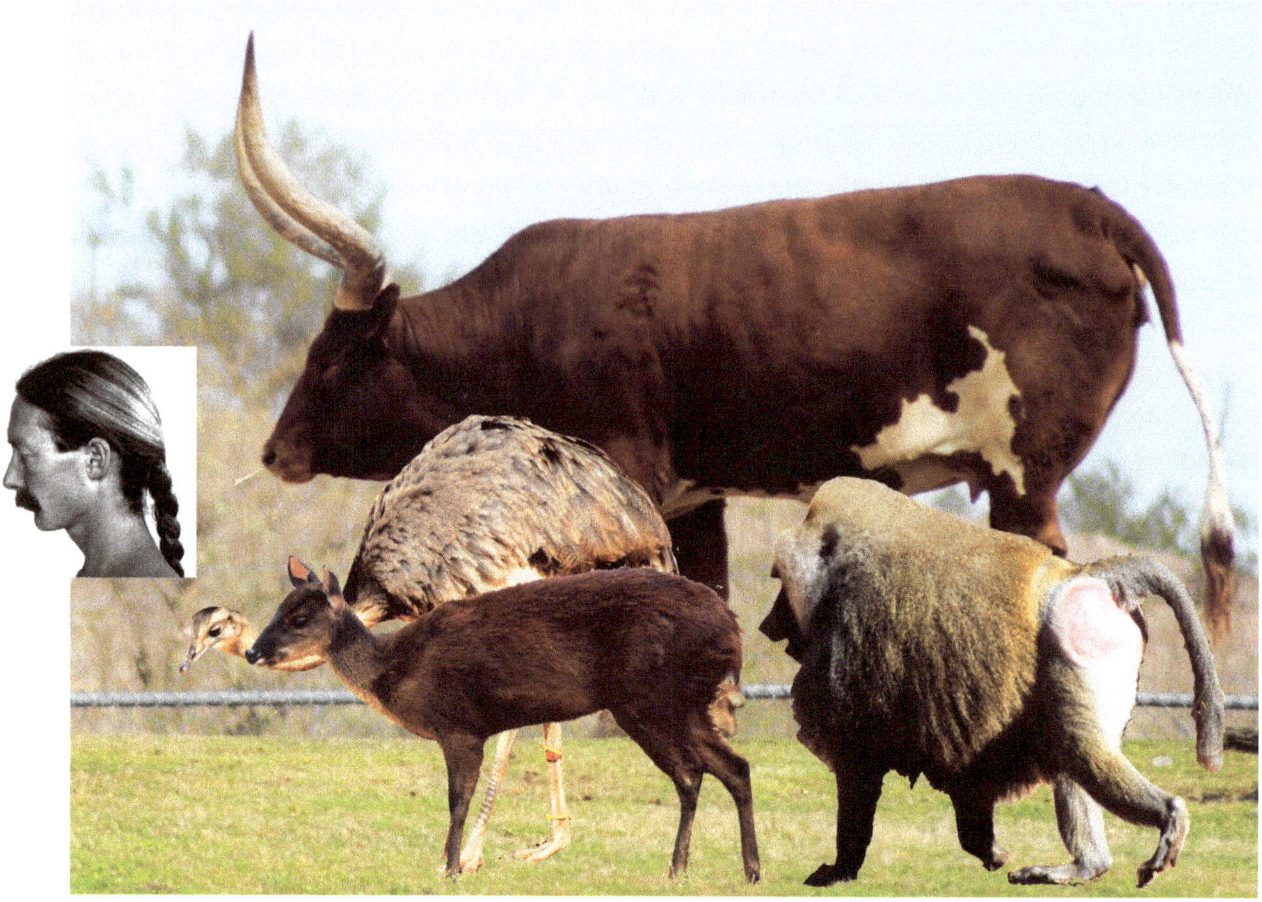

34

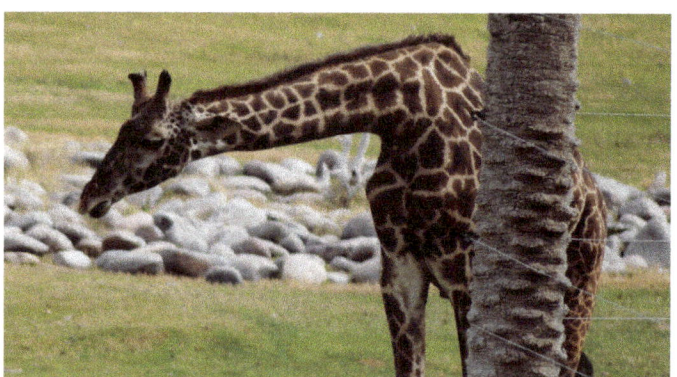

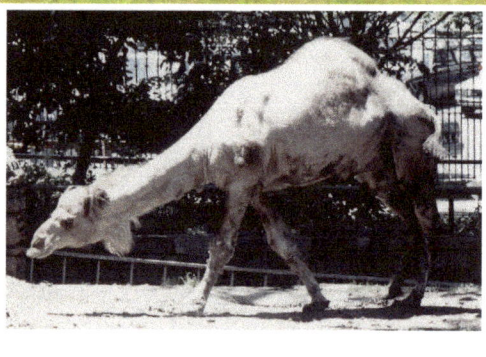

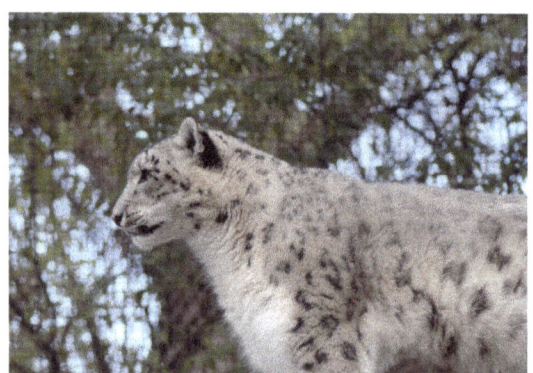

GOING-ALERTLY-FORWARD (GAF)

The primary head position is the going-alertly-forward position. In this attitude, the animal moves or faces forward into space with the sense organs positioned for a balanced array of input. The going-alertly-forward position reads the space primarily in front and near to the animal. The sensory sequence is nose, mouth, eyes, and ears. The eyes are in the middle of the head, top to bottom. The inclination of the nose in these pictures all concur within a few degrees.

HIGH-HEAD-SURVEILLANCE POSITION (HHS)

In the high-head-surveillance position, the animal usually stands still with its head held high, and its sensory input shifted to more distant sight, smell, and hearing. With the head rotated backward, the mouth usually precedes the nose, and the eyes are set high towards the top of the head.

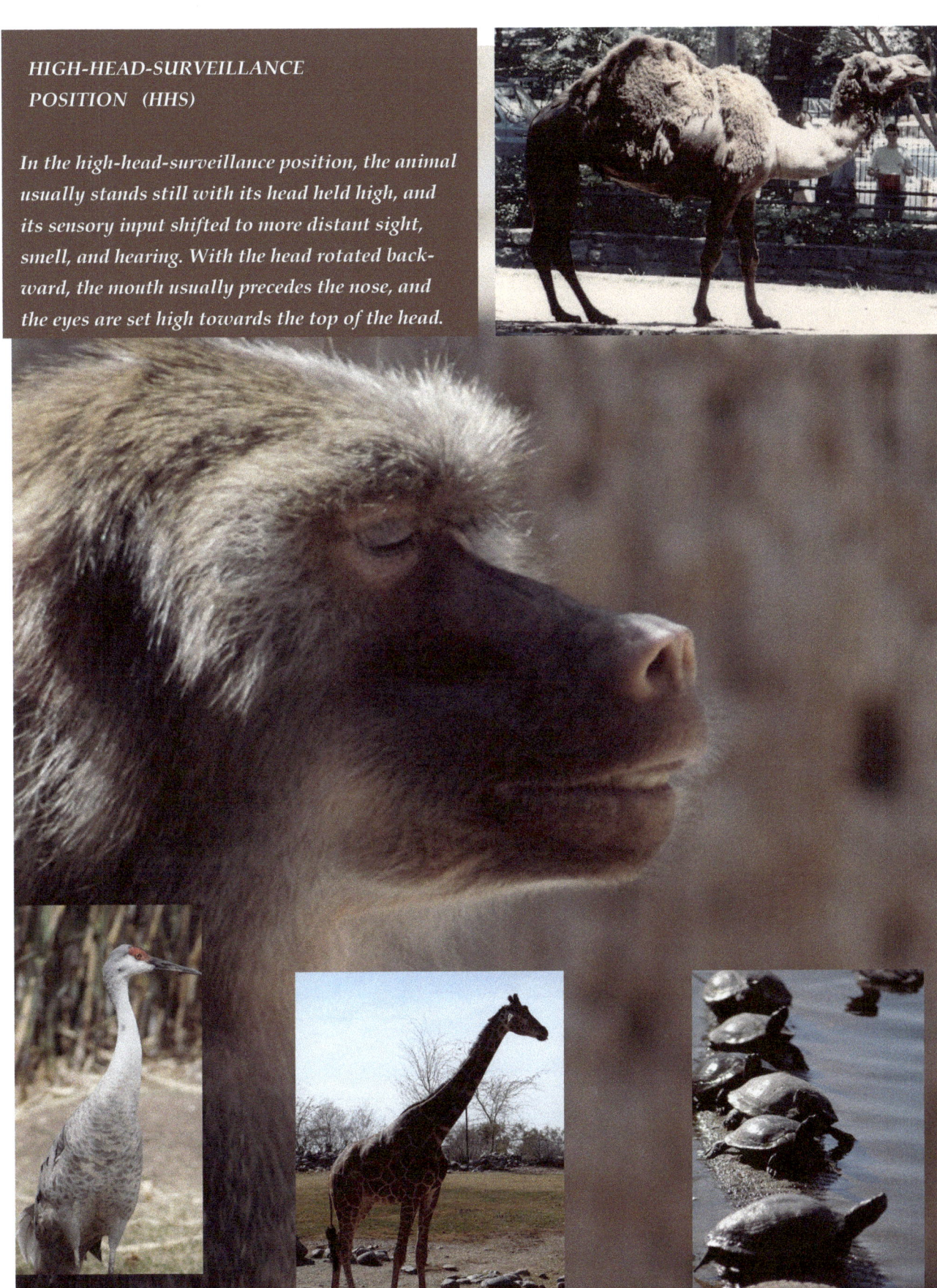

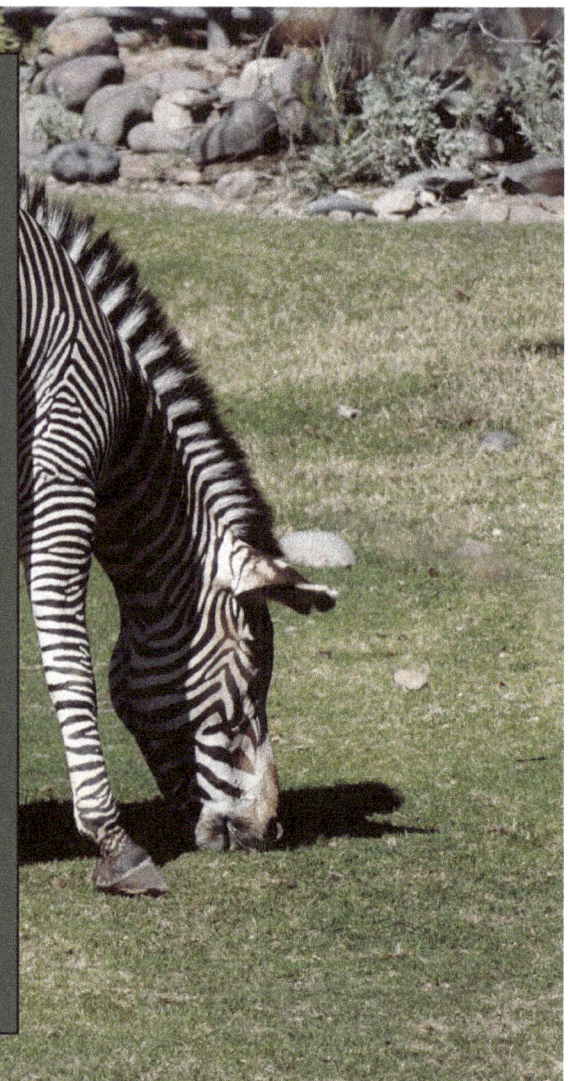

HEAD DOWN FEEDING (HDF)

The head-down-feeding position deprives the animal of much of its sensory input. This is why grazers such, as rabbits or deer, will raise their heads to look around while they chew. Taste and smell are overwhelmed by the process of eating. As the head turns downward the ears rotate to the top and hearing becomes the primary sense. When eating, animal eyes lose much of their focused view to the front but in compensation, have a much greater peripheral field to the rear. You can verify this for yourself by first becoming aware of the limits of your peripheral vision and then bending over at the waist. You will discover, as you bend, your peripheral field of vision into the space behind you increases greatly. The pathway to this enhanced view is down the oral-optical channel. The oral-optical channel is a sight line between the pupil of the eye and the corner of the mouth. The ability to monitor the corner of the mouth is a design feature of land vertebrates heads.

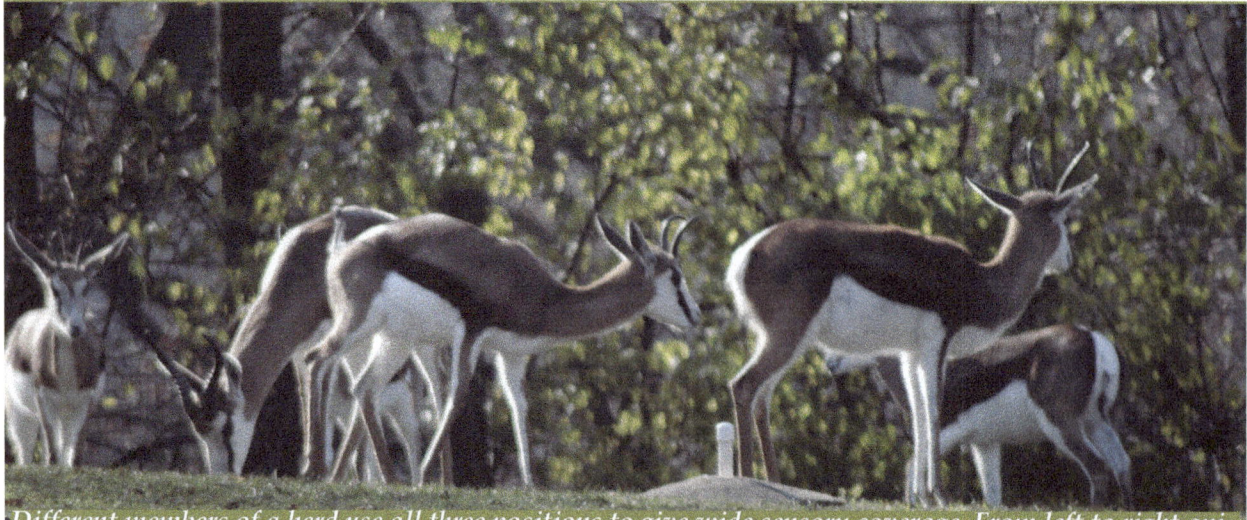

Different members of a herd use all three positions to give wide sensory coverage. From left to right animals are using the moving forward alertly, head down feeding, moving forward alertly and high head surveillance.

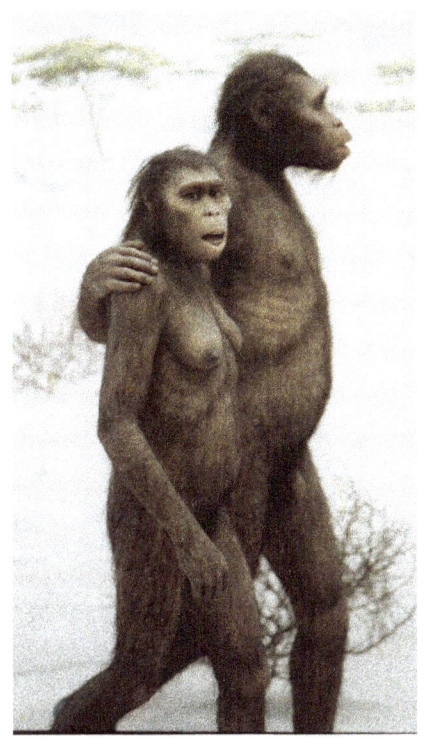

To the left is the Afarensis- Laetoli display at the American Museum of Natural History, New York. The reconstruction is founded on the Table Fallacy. Under that picture is the male's head rotated into the natural Going Forward Alertly position. It is juxtaposed with a young gorilla's which indicates a more horizontal posture than in the museum display. The reconstructed bodies also have attributes of the child like schematic figure which further complicates the situation.

Below is another reconstruction that is even more extreme high eyes linked to the Table Fallacy.

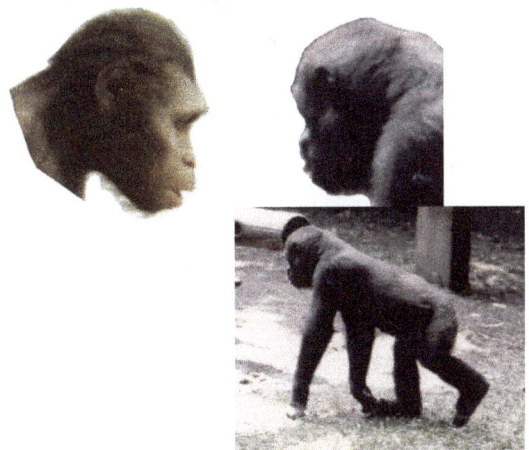

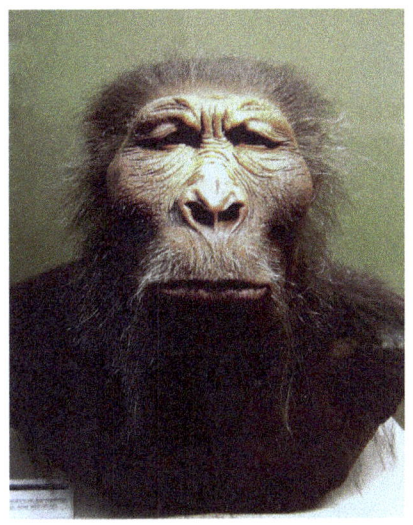

Paranthropus boisei Westfalisches Museum fur Archaologie, Herne
photo credit: Lillyundfreya
Wikimedia Commons

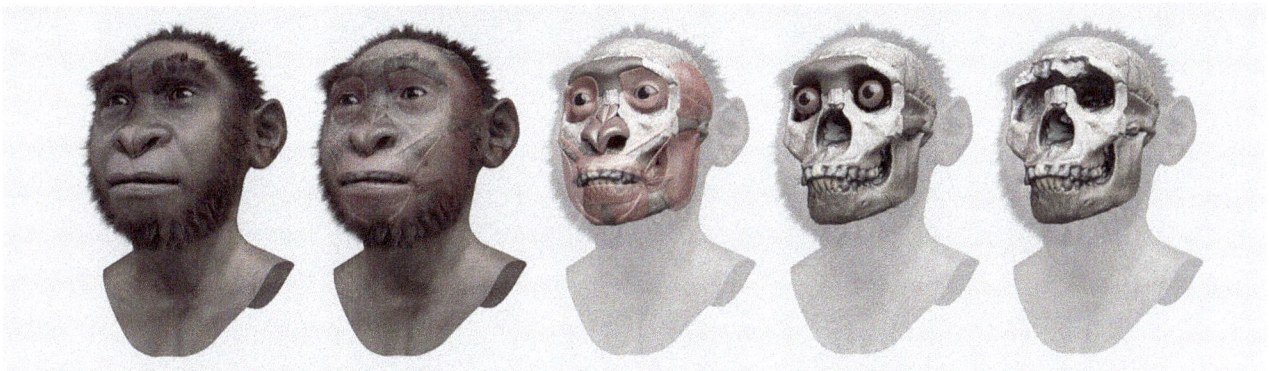

This reconstruction of Tarkana Boy demonstrates the highly subversive effect of faulted imagery. The artist built his work on the table fallacy. Cicero Moraes photo: Wikimedia Commons

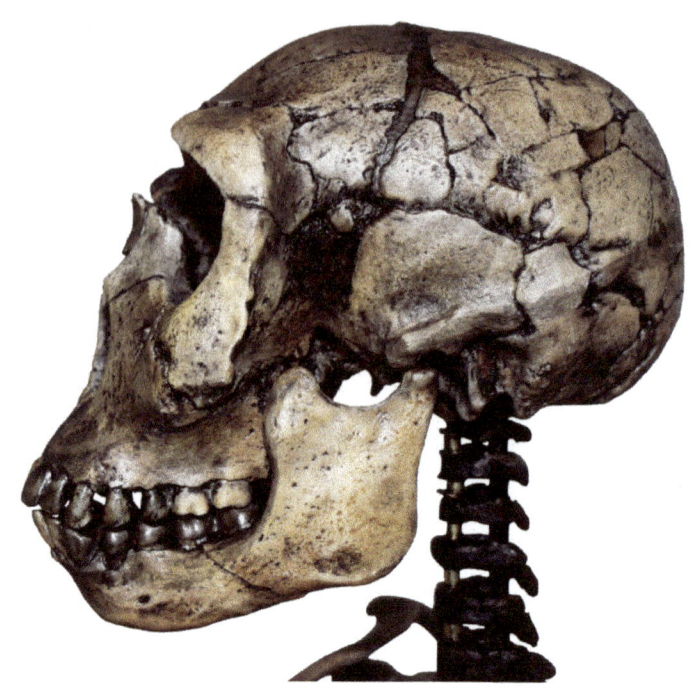

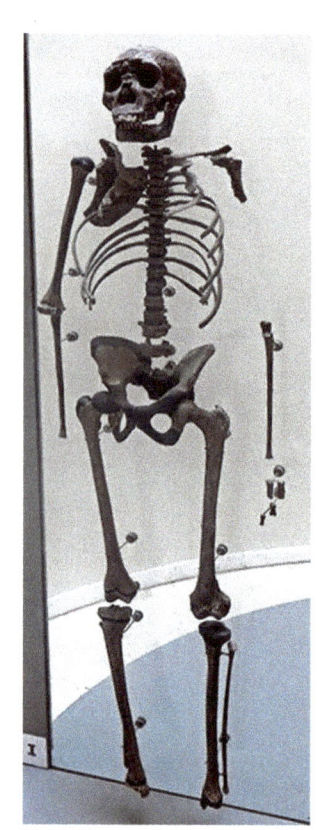

Source: Display at Musée National de Préhistoire, Les Eyzies
Photo: Don Hitchcock 2014

The Turkana Boy skull is placed in the Frankfurt Horizontal position. His projecting lower face, or prognathism, is an artifact of the skull's is positioning. Faces are locked into space, the braincase is rotated in horizontally orientating animals, opening the angle between the facial stalk and the braincase. This causes the jaw to lengthen, stretching from the face to the receding and the ascending mandibular joint in the base of the brain case.

In the full skeleton, the spine is reconstructed to be ramrod straight as is the neck's representation in the skull photo. Our spines have a serpentine curve to balance our vertical figure but the museum version is a projection of the schematic figure.

In the model's small photo on the next page the neck slopes forward. The mastoid muscle, the strong cord that attaches behind the ear, descends vertically to the front of the neck. This places the face, particularly the nose, in front of the body.

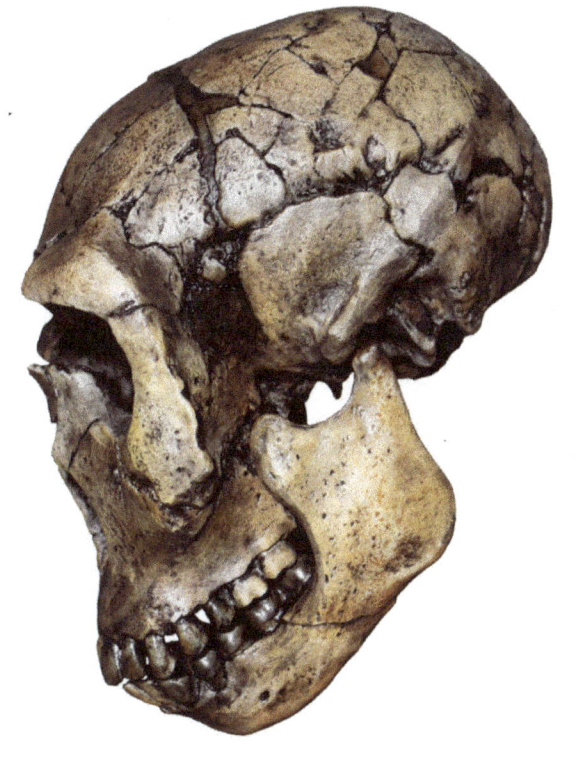

The skull of the Turkana Boy has been rotated to the natural position that it would have held while moving forward. The nuchal line on the back of the skull denotes the contour of the neck muscles. It is an extension of the zygomatic arch as is also true of the skull of the baby gorilla below. In the human skull the nuchal aligns with the mastoid process and rises more steeply. Both the gorilla skull and Turkana Boy have about the same angle of nuchal line rise.

It should be expected that, as in gorillas and humans, Turkana Boy's facial stalk angle would open more during maturation.

The form convergence between the baby gorilla, human, and Turkana Boy is not a result of neoteny but results from the mechanics of vertical orientation.

Note that the very young gorilla is standing up and faces into the environment like humans. The photos of the model and the baby gorilla are included to verify the positioning of the skulls.

Note that the very young gorilla is standing and faces into the environment like humans. The photos of the model and the baby gorilla are included to verify the positioning of the skulls.

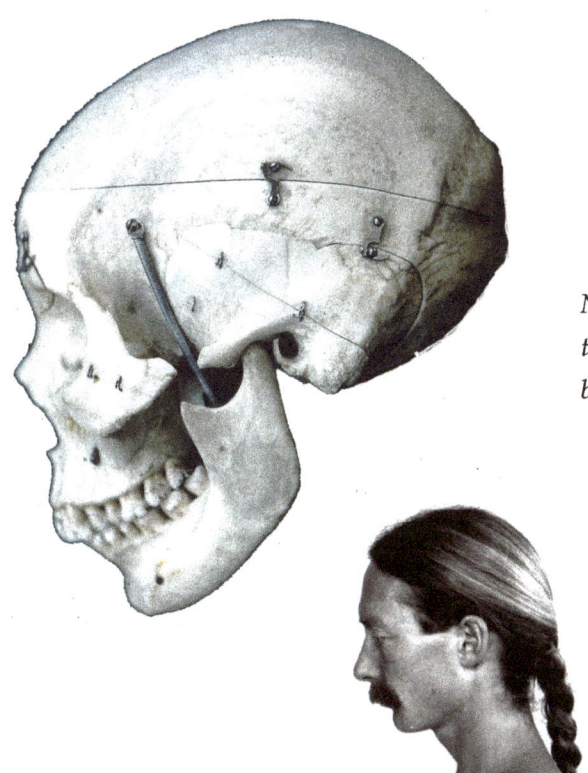

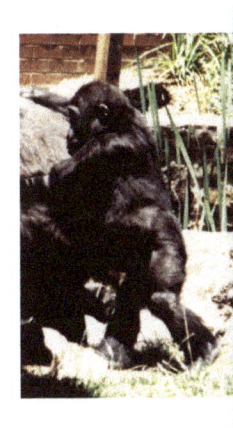

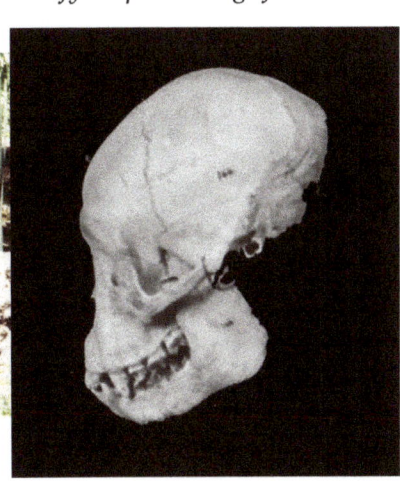

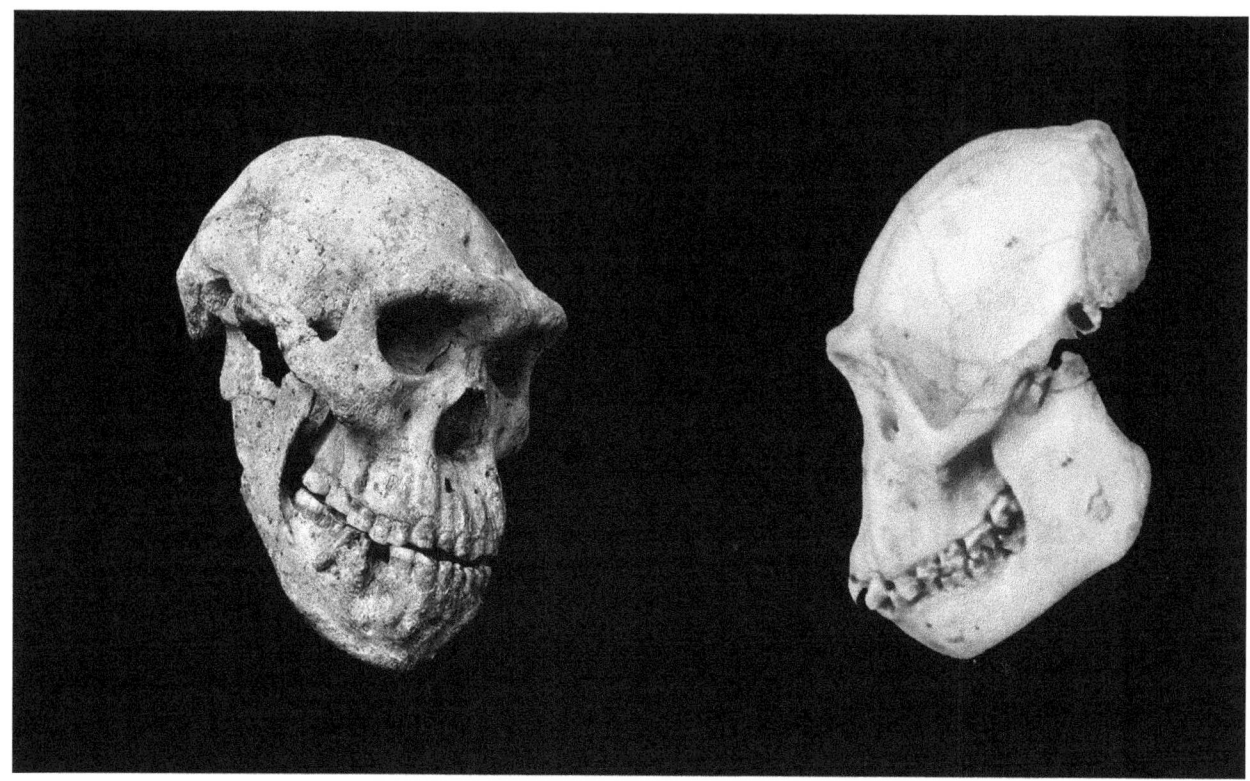

Skull 5 from Dmanisi, Georgia. photo: Guram Bumbiashvili / Georgian National Museum.
Image rotated from the Frankfurt Horizontal to the Going Forward Alertly position.

Adult gorilla from Samuelson's Gorilla Growth Series.

When placed in the going-forward-alertly position offer a cue to how vertically or horizontally the animal was orientating by the position of the ear canal to the bridge nose. In fully vertical humans the ear canal is below the bridge of the nose but in horizontal animals it is above. In the gorilla-growth-series, the ear canal rose with their increasing degree of horizontal positioning as the gorillas matured. The gorilla pictured here is in the middle of the Gorilla Growth Series and appears to be about equal to the Dmanisi skull in this regard.

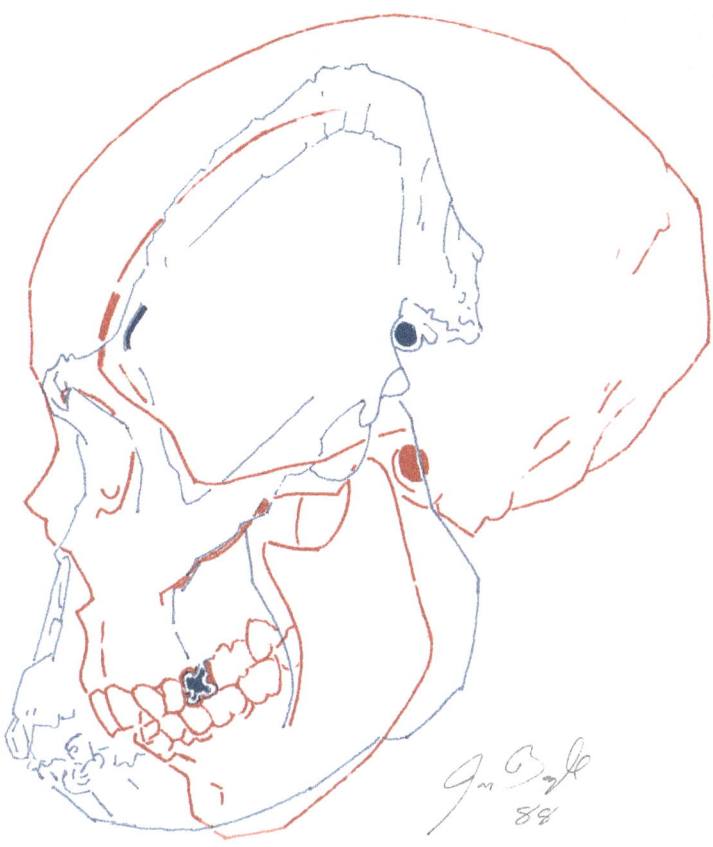

In this comparison, the human lab skull is superimposed on the senior male gorilla from the Samuelson photo. The two skulls are justified at the first molar and the bite plane of the teeth. The human skull was scaled up to match the facial dimensions of the gorilla. This comparison shows the biomechanics of molar pressure containment.

Humans and gorillas last shared a common ancestor eight to nine million years ago yet there is great consistency in the two skulls. The shape and size of the eye sockets and post optical bar are quite similar. The post optical bar is both a strut for pressures and part of the optical field. These competing functions have held the design stable for millions of years.

The margin of the temporalsis muscle on the human skull closely coincides with the sagittal crest of the gorilla, demonstrating that the biomechanics of the two heads are in scale. The lower margins of the two zagomatic arches, where the masseter muscle attaches, match perfectly. The gorilla's zygomatic then arches up following the elevated gorilla brain case. The masseter muscle wraps around the back of the jaw, and the gorilla jaw is squared to give it more surface. The gorilla jaw is more inefficient mechanically being longer with more distance from the fulcrum to the attachment points of the muscles.

Note that the gorilla eyebrow ridges are inside the contours of the human brain case which is why humans don't need them.

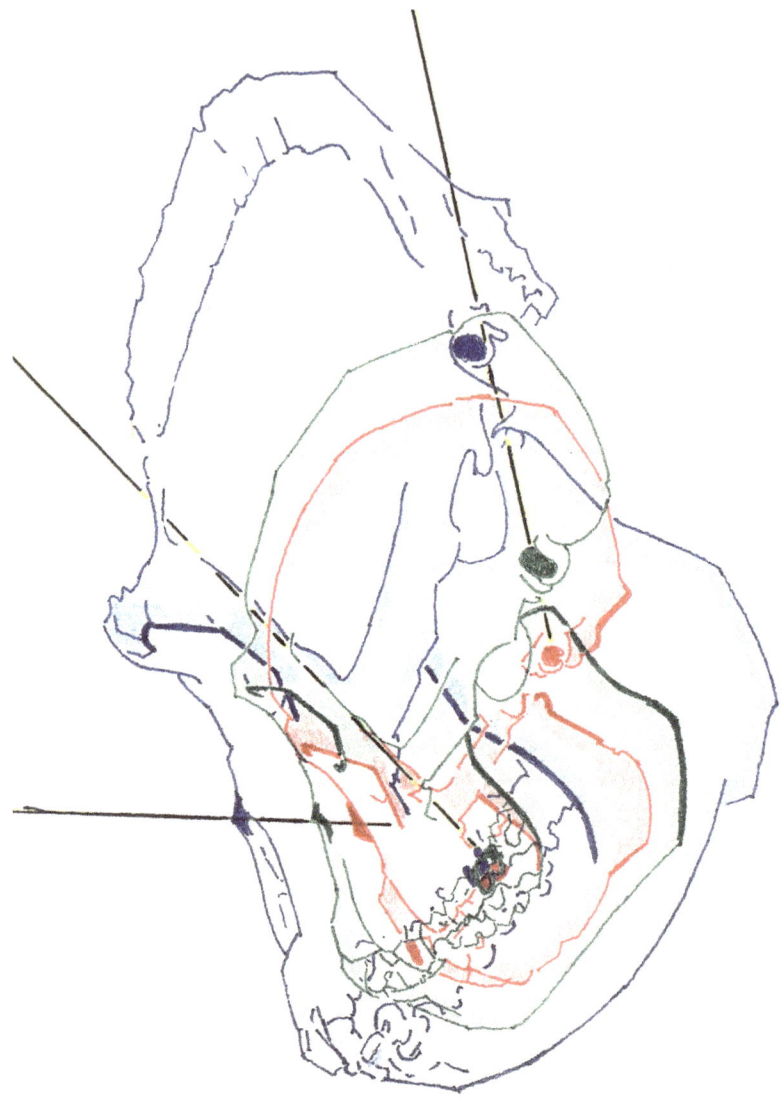

In this image the most senior skull, the middle
skull, and the youngest were aligned using the
bite plane of the teeth and the first molar. This
revealed the structural pattern of the dental
pressure in the skull. Lines connecting the nose,
the ear canals, and the post-optical bar show a
consistent growth pattern.

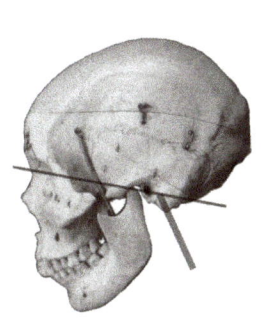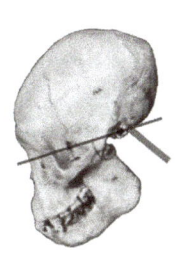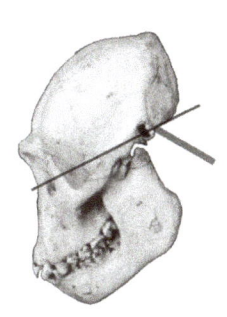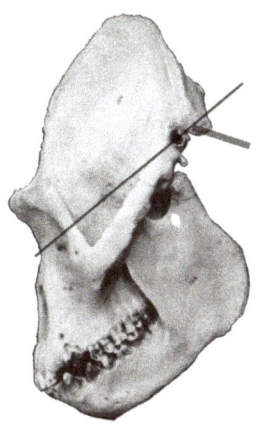

The human lab skull has been added to three skulls from the Gorilla Growth Series. A line drawn from the bridge of the nose through the ear canal on each skull denotes the base of the brain case. A second line indicates the direction of the spine. The spine meets the brain case at a fairly constant angle that is obtuse to the front and acute to the rear. The human skull used here is a single, arbitrary example. Adding additional human skulls from various regions of the world and of various maturations would show a smooth transition from infant human to adult gorilla. This demonstrates a unified design program that adjusts skull mechanics to posture.

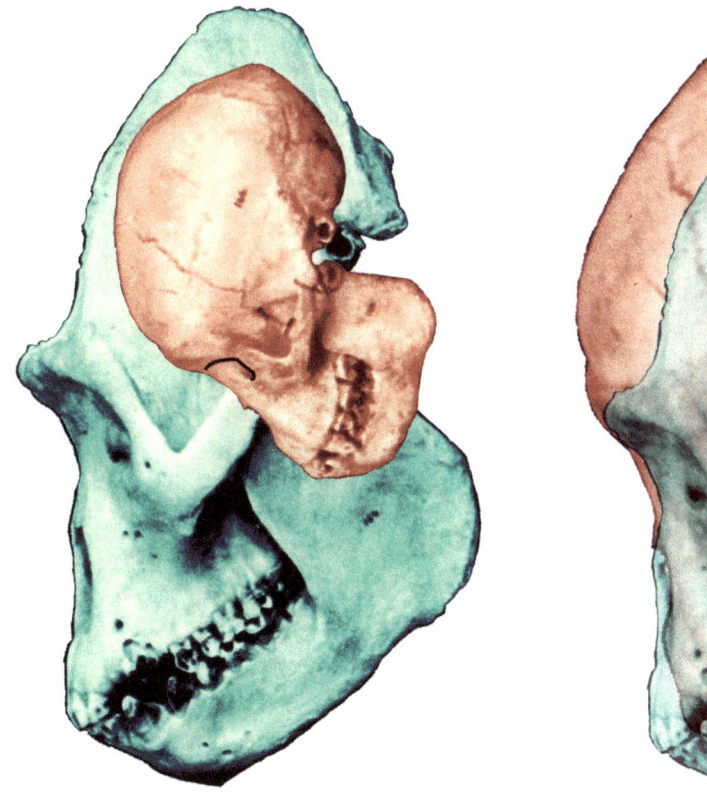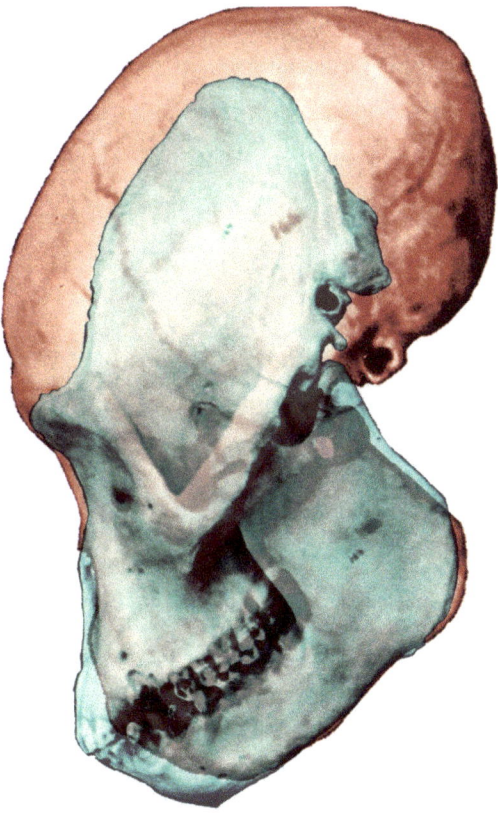

Two comparisons of the youngest and oldest in the Gorilla Growth Series. On the left, both are in scale with the brain cases aligned. In the right comparison the scale has been altered to have the baby's face the size of the adult's and the bite plane and first molar justified. The left comparison indicates a radical reformation during maturation while on the right it appears the structure is little altered during growth. The oldest has been made somewhat transparent and the post optical bar aligns perfectly with the juvenile's as it points into the skull rather than into the eyebrow ridges in the adult.

The hypothetical Horizontal Human Being

Suppose some humans were to take up a horizontal quadrupedal mode of living like the gorillas. How much modification would be needed in the skull to accommodate this new habit? Primate heads are a coherent design set so it is possible to design a horizontally orienting human by adhering to the underlying bio-mechanical formulas. The senior gorilla skull from the Gorilla Growth Series is justified with the Lycoming College lab skull. Both skulls are positioned in their going-forward-alertly position.

The human skull is enlarged so they are not their original size. Enlarging the human skull to match the gorilla's reveals proportional similarities. The two skulls were aligned along the base line of their brain cases, the black line, rotating the lab skull's brain case into its new horizontal mode. The nuchal ridges where the neck muscles attach to the back of the brain case (B) are parallel which confirms the new orientation. The various elements of the human face and jaw have the same shape and proportion as in the gorilla.

The human skull's image was fragmented and the pieces placed to coincide with analogous features in the gorilla's skull. The gorilla skull is rendered somewhat transparently to show the fractured human skull behind it. The darker areas are where additional human bone growth would be needed for the new horizontal mode. This comparison reveals some fascinating relationships that show our close family connection.

The ear canals exactly coincide along with the hinge joints of the jaws. The rear section of the human jaw fits almost perfectly into the outline of the gorilla's jaw. The bite plane of the teeth were matched and the first molars overlain (E). This match then organized the elements carrying the pressure from the molar region. The zyagomatic process and the post optical bar are matched in shape and proportionate size . The underside of the zyagomatic complexes, where the masseter jaw muscles attach, match perfectly (D) as does the leading edge of the ramus of the jaws. The human skull has the margin of the temporalis muscle marked with a dotted line which corresponds closely with the gorilla's sagittal cress, the boundary of its temporalis muscle (A).

Matching the images along the bite plane at the canine teeth shows us where the gorilla has been hiding its chin (F). Gorillas are accused of not having a chin which is thought of as a uniquely human attribute. Both our chins, and theirs, buttress the lower canine teeth (see page 92). While the previous discussion of dental forces focused on the pressures

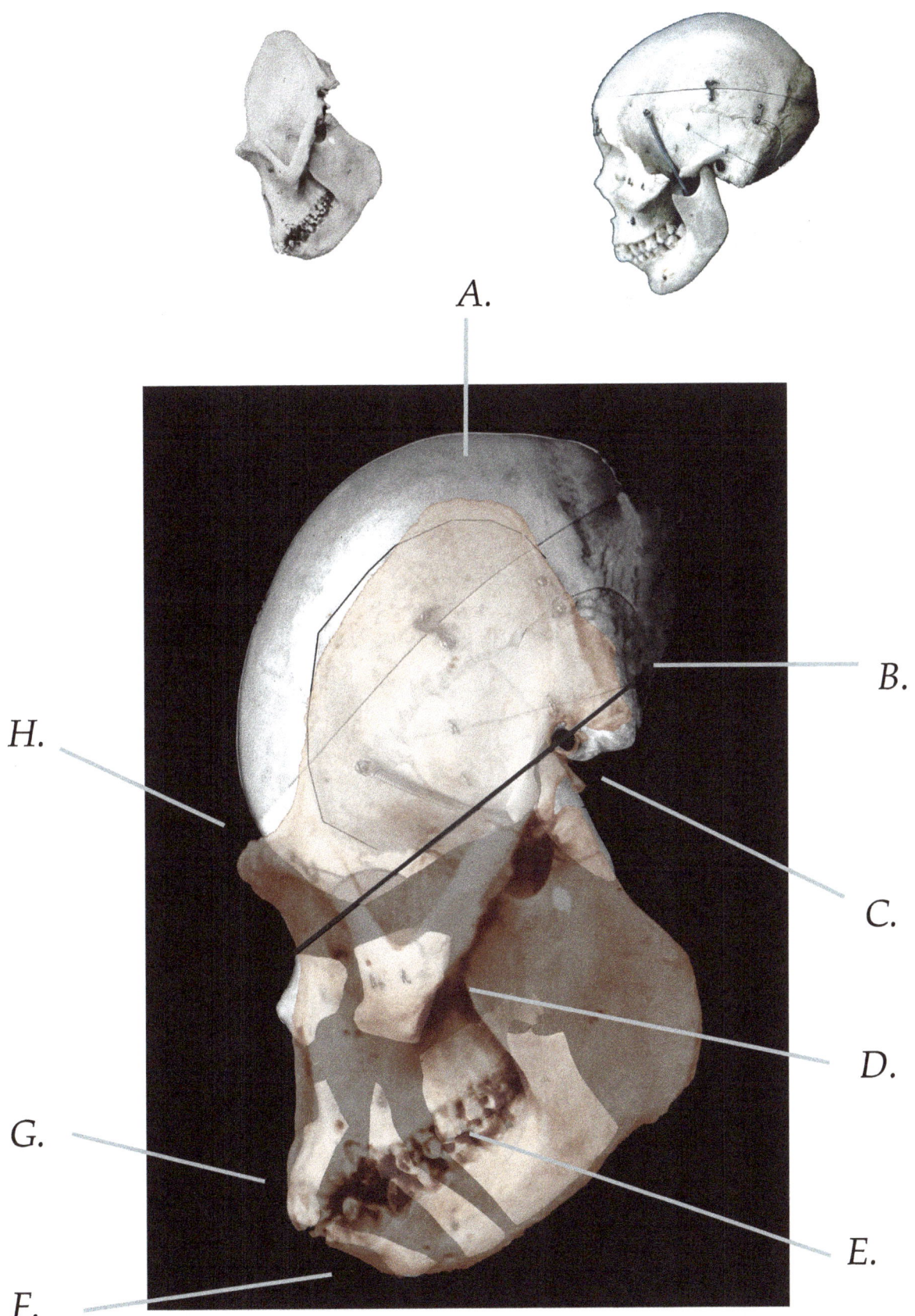

A.

B.

C.

D.

E.

F.

G.

H.

from the molar region traveling up the outside edge of the eye socket, a second line of support receives the pressures of the canine teeth.

Like the molars, the canines generate a great deal of force, more so in our primate relatives with large fighting fangs than in us. These forces are carried up the walls of the nasal cavity on either side of the nose (G). This is the anvil part of the equation. The hammer is the chin. The chin is obvious in humans because our jaws are short and our front teeth come together vertically putting the canine's buttress (chin) out in front.

Unlike in human's vertical incisors, the gorilla's upper and lower incisors jut forward and meet at a considerable angle (F+G). This can be seen in the gorilla's image where the human teeth have been overlain. It is possible that this is to achieve the prescribed facial entry into space with less investment in bone growth.

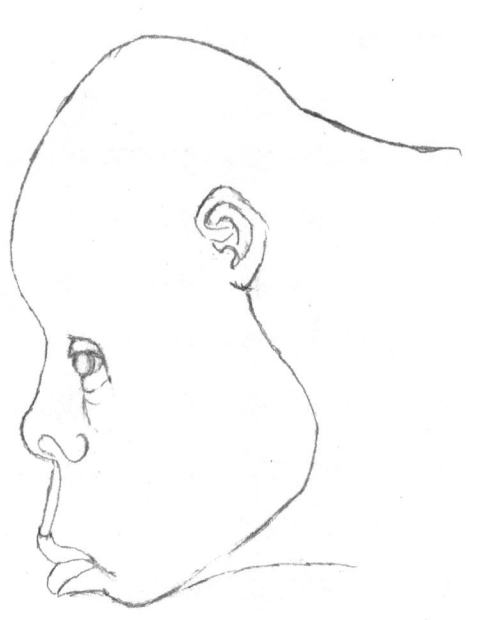

In this reconstruction, we arrive at a hypothetical horizontal cousin that is no less human than we are. Our small delicate face is an attribute of our vertical stance not our humanity. Our new friend would still have a large brain and be a forward looking fellow, though with a somewhat lowered horizon. Finding an occupation for the hypothetical-horizontal- human would be a challenge. He would be a perfect football lineman, a tackle or guard. His low center of gravity and four point stance would permit him to drive forward using four limbs all the while seeing where he is going. Other sports or positions would be unlikely because the ability to throw would be quite limited. Vertical humans harness the power to throw, which is like an energy wave, starting from the toes, then flowing through the legs, pelvis, torso, shoulders, arms, and fingers. Perhaps he could find a job inspecting the interior of large pipes.

THE FRANKFURT HORIZONTAL

The Frankfurt plane was established at the World Congress on Anthropology in Frankfurt am Main, Germany in 1884, and decreed as the anatomical position of the human skull. It was decided that a plane passing through the inferior margin of the left orbit (the point called the left orbitale) and the upper margin of each ear canal or external auditory meatus, a point called the porion, was most nearly parallel to the surface of the earth at the position the head is normally carried in the living subject. Wikipedia--Frankfurt plane

You can take the Frankfurt Horizontal for a test drive. Raise you head so your eyes are fixed on the horizon which is always at eye level. Now go for a walk but be careful, the Frankfurt Horizontal doesn't give you a visual check on your footing and worse things can happen than stepping on the cat. When we are normally walking we drop our gaze to check our footing. The more treacherous the footing the lower the gaze. While in Nepal, I went up into the mountains and remember being quite frustrated. I was in this astonishingly beautiful place but I kept looking at my feet as I walked. I wanted to see the landscape but I was afraid of joining it.

Over the last few decades the Frankfurt Horizontal has slowly emerged for positioning skulls in a "lifelike" position. This is an attempt to find a universal position from which to compare skulls. Paleontologists have been quite up front in admitting that they have difficulty finding a good place to start a system of comparison. The Frankfurt Horizontal system uses two points on the skull as landmarks. The lowest edge of the eye socket is placed level with the highest point of the bony projection of the ear canal. In most humans this position sets our heads as if we were looking out to the horizon. Aiello and Dean's <u>Human Evolutionary Anatomy,</u> page 210: "All primates, as well as many other animals, habitually hold their heads so their eyes look forward towards the horizon." In Johanson and Edward's <u>From Lucy to Language</u> p. 83, we get this about the Frankfurt Horizontal: "This position replicates the natural angle of the human head." The Frankfurt Horizontal coincides with a position in my system of analysis: the high-head-surveillance position, where we would look up and out to be alert for things occurring at a distance. We don't normally walk around like this or we would be tripping over things. The thing about this quote in <u>From Lucy to Language</u> is that it is printed opposite from a full page photo of a reconstruction of a male Australopithecus Afarensis by artist John Gurche. Gurche, who has spent a career hanging around paleontologists, undoubtedly knows about the Frankfurt Horizontal. Gurche's Afarensis is photographed as looking directly into our eyes, aligning its face just as we do. In this position, it wears its ears above its eyes, like a gorilla's, giving some disquiet to Johanson's and Edgar's contention that Afarensis was predominately bipedal. The geometry of the head is directly linked to the animal's orientation from horizontal to vertical during its life. Gurche's reconstruction, as photographed, appears to be on the money, and shows a more horizontally orientated primate. In the expanded and updated version of From Lucy to Language (2006) Gurche's reconstruction was replaced by a skeleton of a neanderthal with the skull in the Table Fallacy position. No doubt that substitution has nothing to do with the issues I have raised here.

Interestingly, in describing the Neanderthal skeleton the authors equate some of its features with that of living human populations. Its shorter limbs, stouter body proportions, and larger cranial capacity like those of northern living populations: the Lapps, Saame, and Greenland Inuit. Further down on the same page the authors contend that these differences along with others, show that the Neanderthal were a separate species not a subspecies of modern humans. In just a few paragraphs the authors made the argument that form follows function and then the opposing idea that inheritance dictates form.

Finding fossils may be a great outdoor adventure but analysing them has always been an indoor activity. Those making the comparisons often seem unwilling to test their conclusions against external reality. They don't make the trip to the zoo to see if a particular prognathous animal actually has trouble seeing in a binocular way. They never stand up and walk around the room to test their conclusions about how a bipedal animal strides. Perhaps we should expect that, having proclaimed that all primates always hold their heads in a certain manner. The researcher should take a short walk with their own primate head in that position to see if it actually works. The thought that conclusions should be tested doesn't seem to occur very often.

The Frankfurt Horizontal distorts the analysis in humans because human heads are tuned to the visual environment they inhabit. People who live in open places, deserts and ice fields, look up to the horizon where opportunity and danger is most likely to be first seen. For most people this will rotate the mouth up and in front of the nose which offends the sensory design rule. If this becomes habitual, the faces in that group will rotate farther back under the brain case.

EYE LEVEL

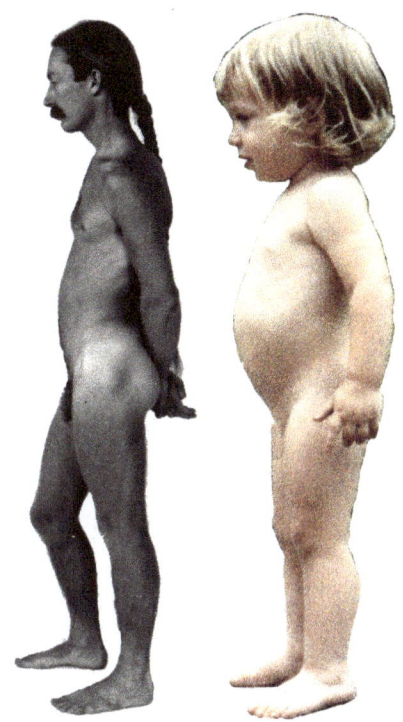

During our lives our facial stalk rotates outward. When we are young our heads are large in proportion to our bodies so the head can be on top and the nose still meets the ancient design parameter of being in front of the body. As we mature our bodies grow more than our heads so we lean our heads forward to keep the nose in front. The facial stalk rotates outward to maintain the profile and the lower jaw grows longer to keep up. Around our twentieth year, the jaws, upper and lower, are long enough to accommodate the third molars, the wisdom teeth. Unless, of course, during the teenage years the family dentist declared them impacted and had them removed.

Groups of people that live in heavily vegetated surroundings have their long distance vision blocked so they lower their gaze to be aware of dangers nearby. To compensate, their faces rotate outward. Each group, in their home location, is placing their face into space in exactly the same angle but have had to modify their head structure to do it.

The contours of the skulls of various peoples around the world are very consistent as are

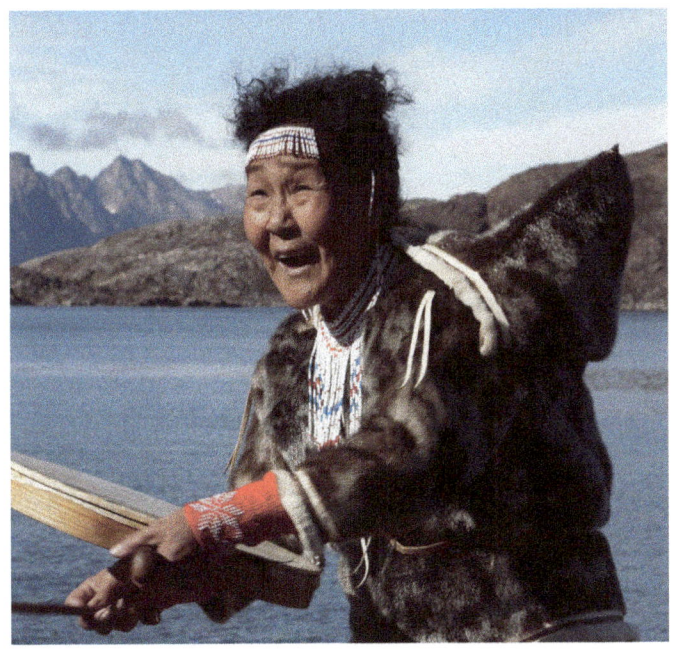

shaman dancing with tabour, Greenland, Ammassalik, Alamy

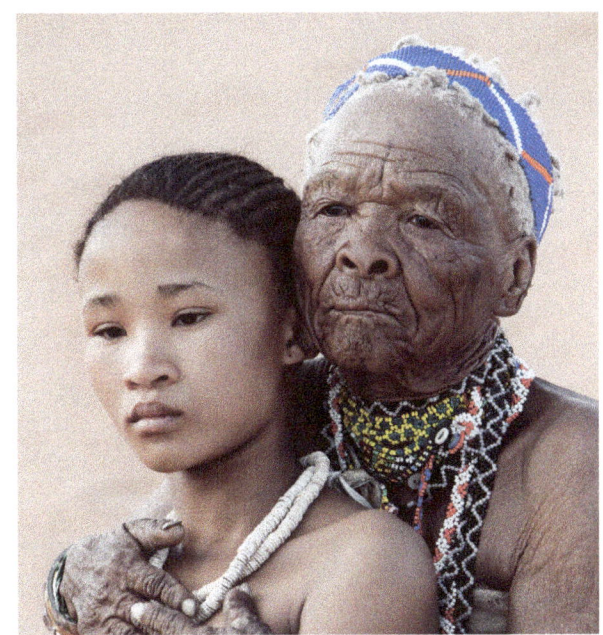

San girl and grandmother Alamy

their large frontal lobes. Opening the angle of the facial stalk to the brain case floor pushes the eye sockets upward in the brain case giving the appearance of a sloping forehead.

When the facial stalk is pushed back in open visual environments, there seems to be two ways: a minor reform and a major reform. The San who live in the Kalahari Desert in southern Africa and the Inuit have the major reform. Their facial stalks have compressed back into and under the skull. This seems to also be the structure of Asian heads but, of course, Asians are a very large population and live in a great variety of environments. I suspect that the basic Asian plan shifts to fit diverse visual environments. A study would be needed to identify people of the same genetic group living in visually diverse locations.

Another way of dealing with an open visual environment is to rotate the lower part of the face rearward. The rotation begins at the bridge of the nose with the shortening jaw taking the mouth rearward. This pulls the face away from the nose which then becomes prominent. This reformation is less comprehensive than seen in the Inuit where the entire facial stalk rotates.

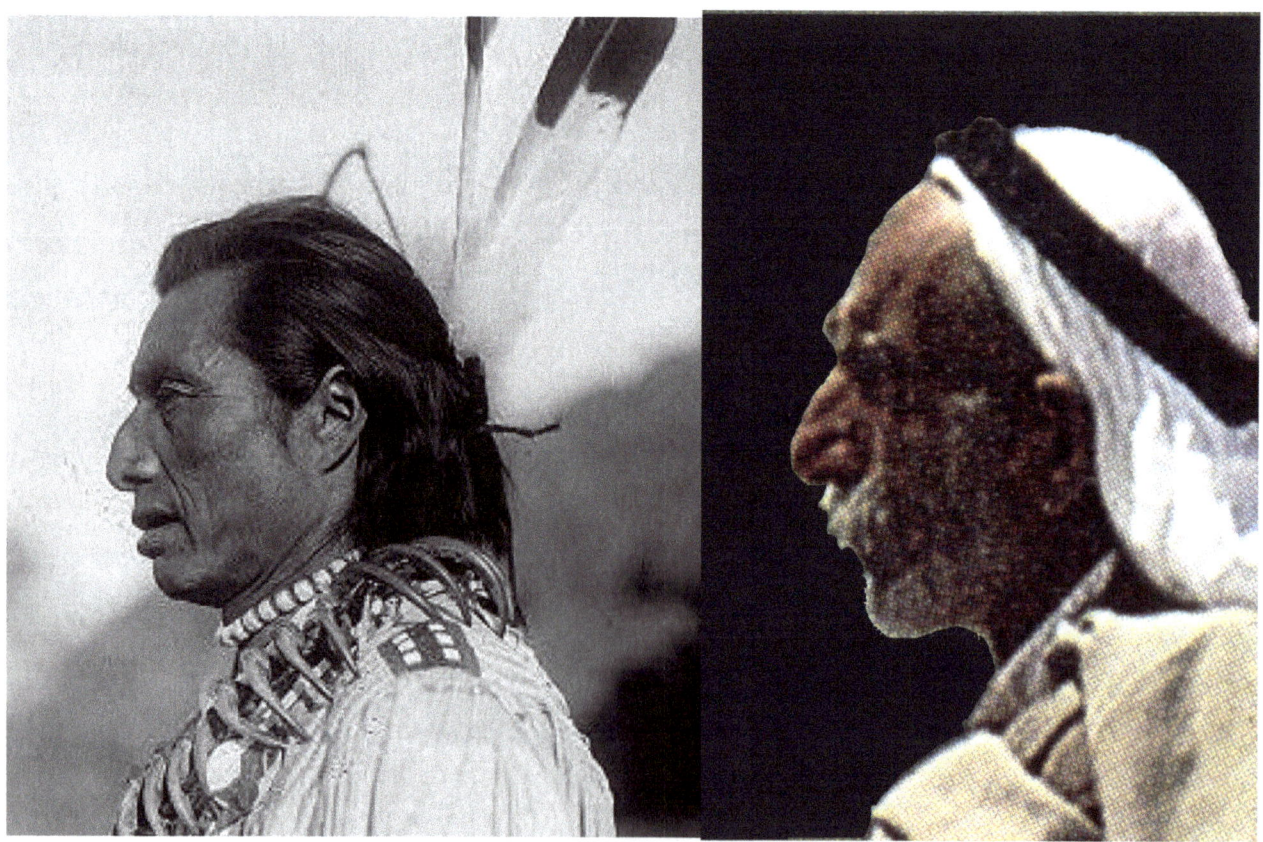

Crow's Heart, a Mandan, wearing a traditional deerhide tunic, photo by Edward Curtis, ca. 1908. (detail) Wikimedia Commons and a traditional Arab herder.

When the giant Olmec Heads were first discovered in the late nineteenth century, they were thought to have an Africa connection. Books were written espousing this theory. Recent genetic analysis from Olmec burials show that they were a native American people.

Olmec was the first and oldest civilization in Central America. The heads along with other figurative art shows that they were primarily a trading culture. This is supported by their trade goods being found over a large geographic region.

The Olmec head, although two to almost three thousand years older than the Benin one below, is much more naturalistically observed. Naturalism in art is an attribute of a trading culture.

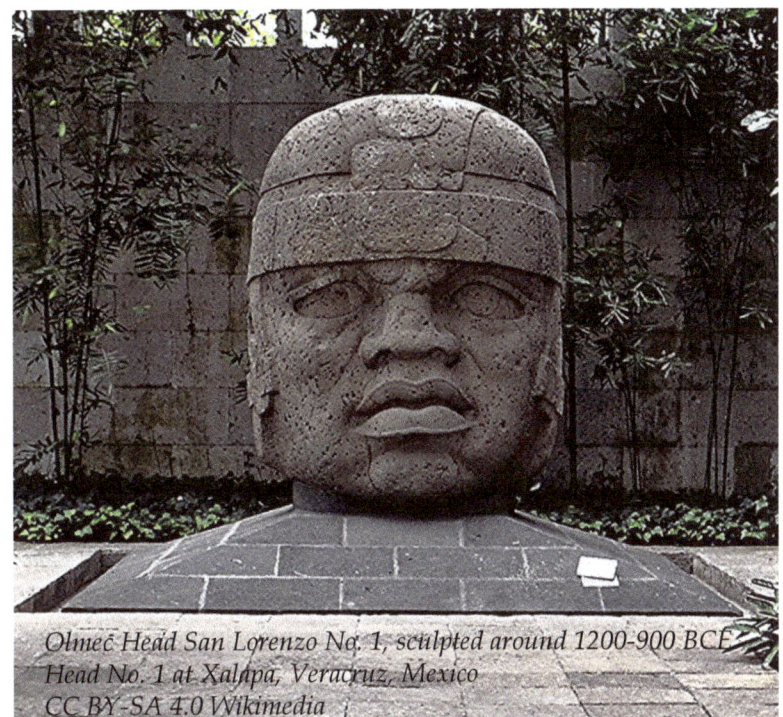

Olmec Head San Lorenzo No. 1, sculpted around 1200-900 BCE Head No. 1 at Xalapa, Veracruz, Mexico CC BY-SA 4.0 Wikimedia

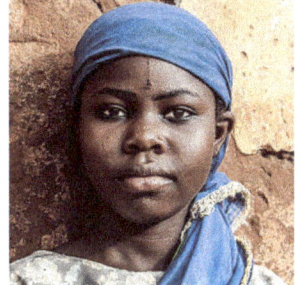

Istockphoto

Both the sculpture heads and that of the little girl above are adapted for an enclosed visual environment. Both the Olmecs and Benins lived in a dense tropical forest.

Seeing the Olmec art heads researchers expected them to be dark skinned and therefore represented a lost people. People with Olmec features still live in the area but are not dark skinned.

By necessity, people living in dense visual environments drop their gaze neared their feet. Their facial stalks adjust by rotating outward to comply with the ancient protocol for facing correctly. See the comparison between the Inuit and Zaire skulls page 51.

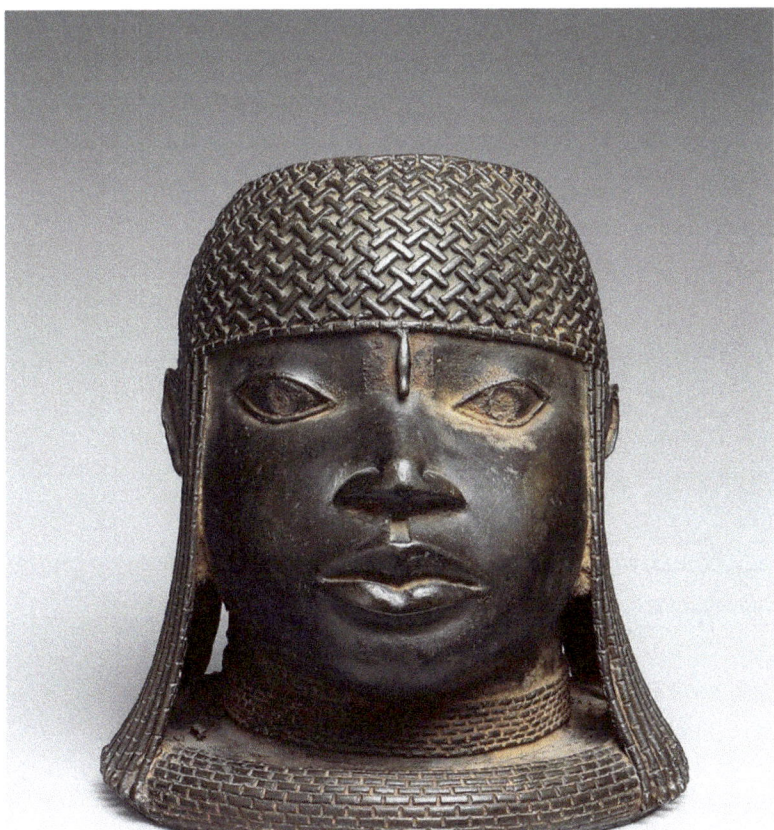

Head of an Oba : Edo artist 16th century Nigeria, Igun-Eronmwen guild, Court of Benin Metropolitan Museum, New York

LOW BUDGET GORILLA

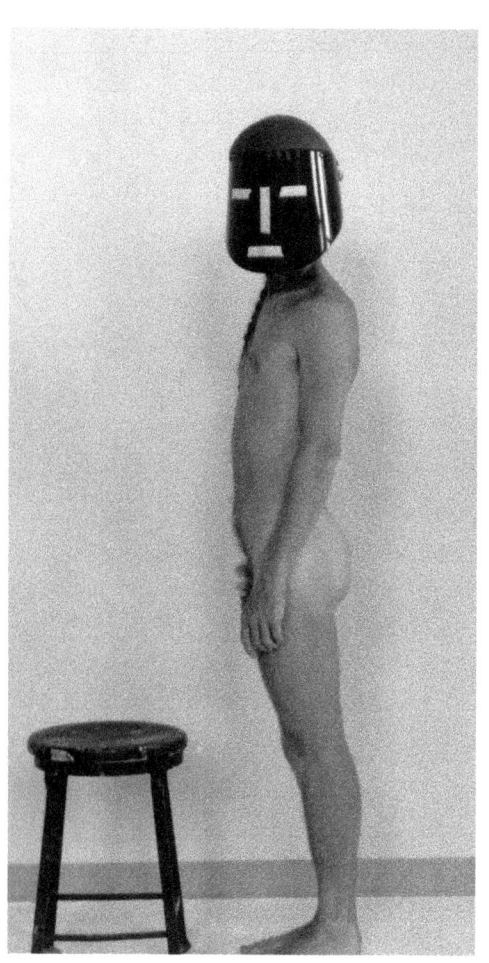

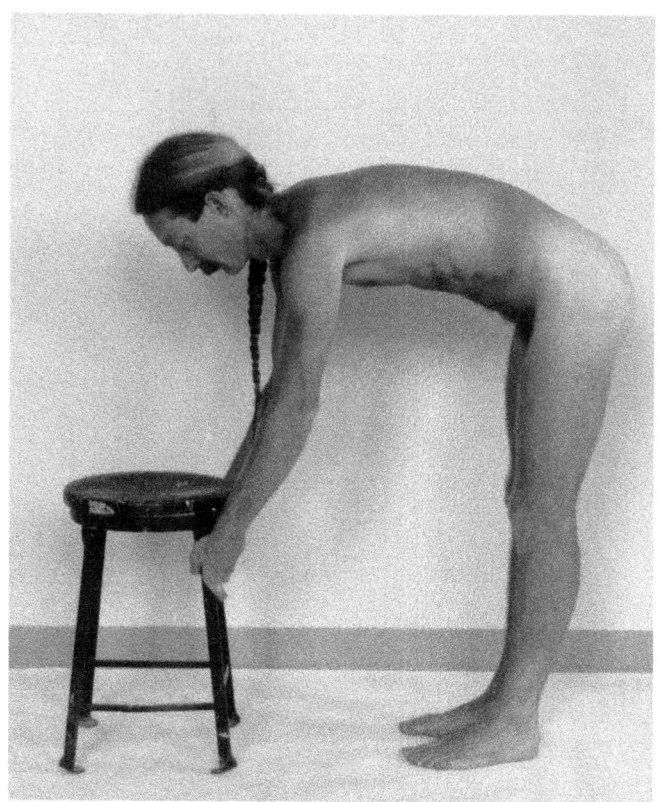

The model was posed with his back horizontal, like that of a gorilla. In this pose the ears are higher than the eyes and they are turned downward which coincides with how they are in a gorilla's head. The external ear and ear canal move with the brain case. Bent over with his spine horizontal, the model finds it difficult to see forward.

In creating the Low Budget Gorilla I wanted to demonstrate two things. As the model bent over to approximate the posture of the gorilla his face would have to rotate outward to operate correctly. How a gorilla or ground dwelling primate turn their heads is different from humans.

The model was equipped with a new face that could be pulled out when he went into the horizontal position. This welder's mask, which hinges out at the temples, and a bit of tape for features, fit the bill.

Here, the model is standing with his new mask face turning towards us normally by swiveling the vertebras of his neck. When he bends to become horizontal in the next page, the mask swings outward to give his new face the proper orientation.

,

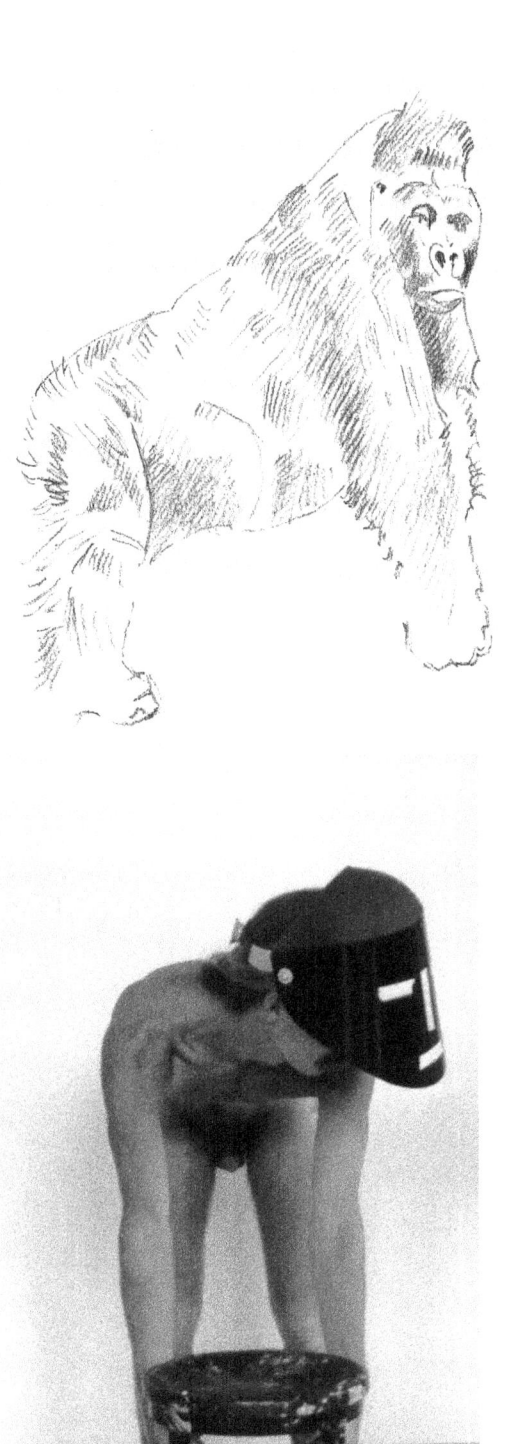

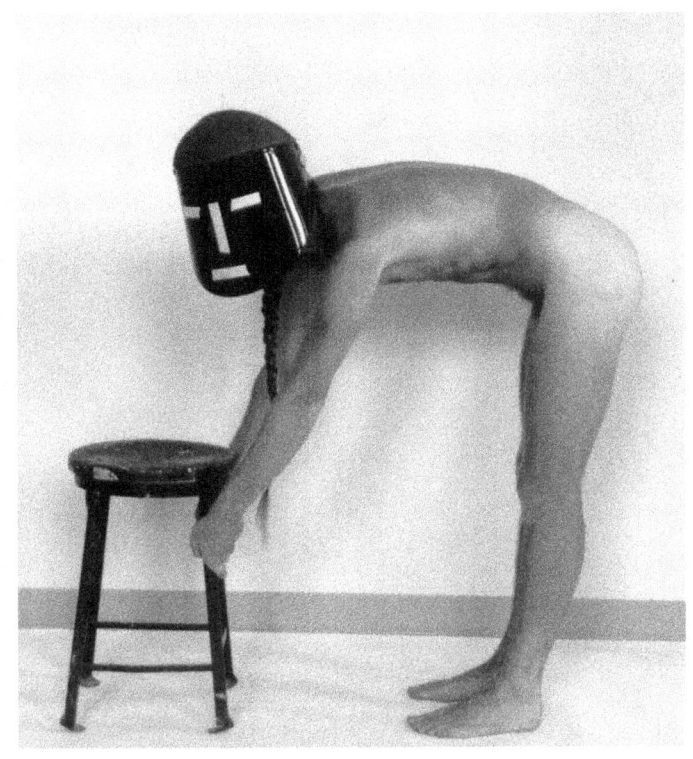

Turning to "look" with his new welding-mask-face, the model is using the same mechanism a gorilla would use. Instead of swiveling the neck vertebrae by using the mastoids, he now curves his spine using the neck muscles that attach to the back of the skull to pull his "face" from side to side. The motion is the same one we would use if we leaned our heads to touch our ear to the shoulder.

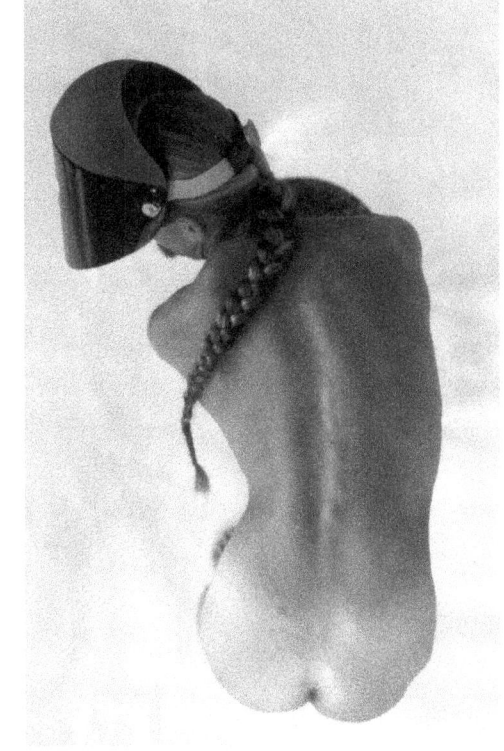

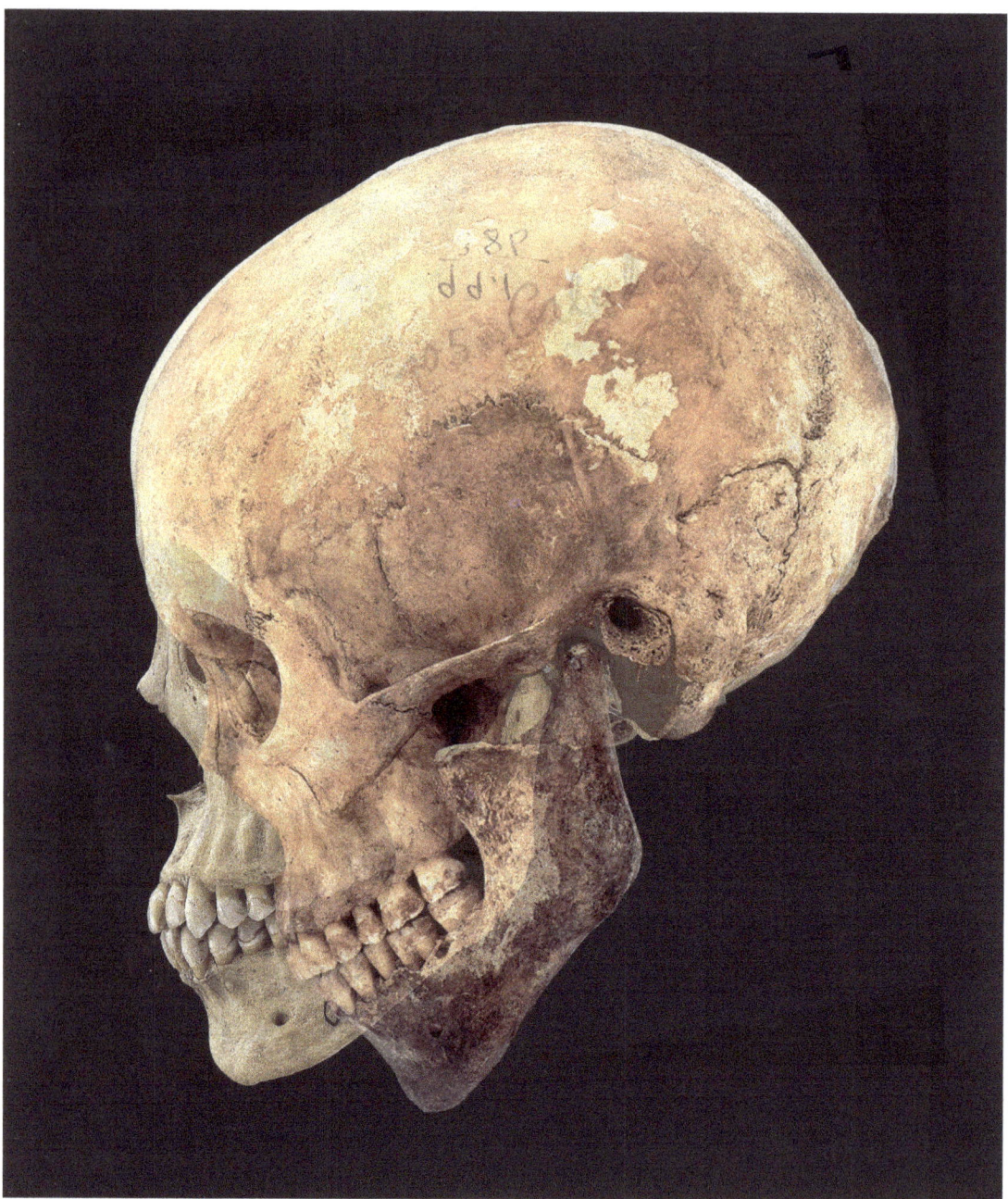

Photos: David Brill

These two David Brill photos are an Inuit skull superimposed on the Zaire (Congo) skull with both brain cases justi-fied. The two brain cases match up almost perfectly which can be seen in the sliver showing at the edge of the brain case contour. The top Inuit skull has been made somewhat transparent for ease of comparison. The position of the facial stalk reflects an adjustment to their contrasting visual environments. The Inuit's facial stalk has compressed under the braincase and has even pushed the ear canal back in the skull's architecture which is unique to this skull. Zaire's facial stalk is angled forward and its ear canal is in the normal position and is located under the jaw socket of the Inuit. Note that the Zaire's eye socket is visible under the Inuit's forehead. The growth or absorption of the frontal skull bone is the driver that swings the facial stalk inward or outward and relocates the eye socket in the brain case.. The front portion of the brain case floor is also the roof of the facial stalk and moves with the facial stalk as it adjust to visual conditions and maturing. The divergent plane of the floor of the human brain case is not a factor of being human but of being bipedal.

STANDING AROUND

Before we can walk we need to stand and the mechanisms in standing are helpful in understanding how we walk. The normal standing posture is the classic contrapposto pose. You will find this pose, with only minor variations, in Classical and Renaissance figurative sculptures or whenever you find folks hanging out on a corner.

Sculptors from the classical age on have loved and used this pose because it is very

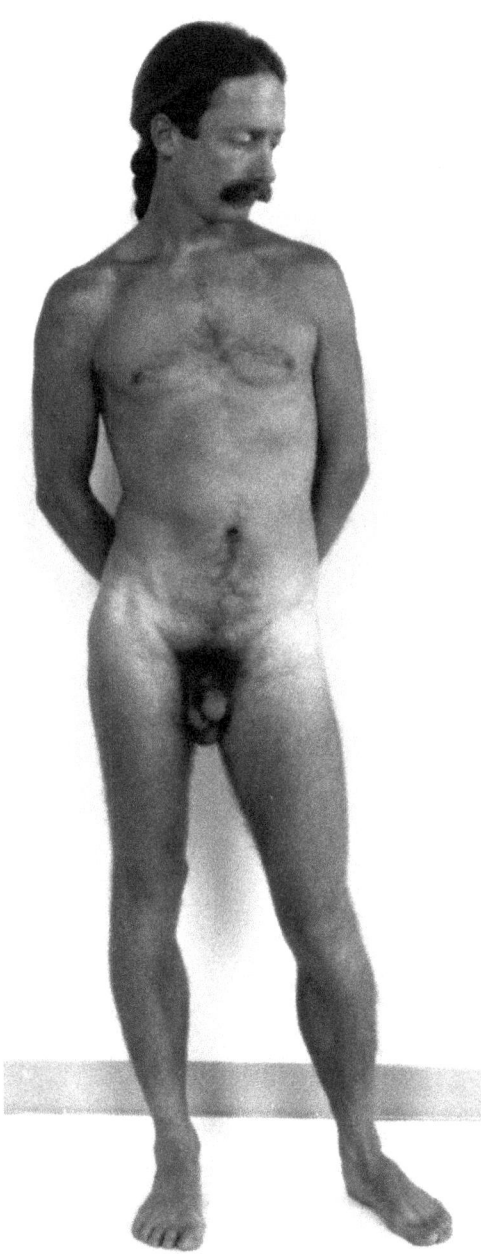

characteristic and real while at the same time having a wonderful flow and logic. Ancient Greek sculptors, like classical philosophers and mathematicians, desired to understand the world factually. Along with their contemporaries, they studied the figure logically and analytically. Breaking it into three essential questions: what is the function of the structure, where and how can it move, and where is it immobile?

The function of the standing figure is to balance in space. Since we go about on two feet, we are required to vertically stack our body carefully around its center of gravity so that the weight terminates within the area of the feet. Standing is a bit precarious since most of the weight is usually on one foot with the center of gravity being near or in the area of that foot.

Stand with your two feet a little apart, and with your weight equally balanced. Rock forward and feel the pressure shift onto the toes and the ball of your feet. Rock back and feel the pressure shift to the heels. Normally when we stand we don't feel a specific location for the weight within the feet so we know that the weight is equally distributed to front and back. Simple deduction tells us that we are taking all the downward pressure of our bodies on the arch of the feet and that the arch is distributing it equally, front and back. It makes sense that the weight would be supported on an arch form. In a standing pose the center of gravity is near the arch of the foot that is receiving most of the weight.

Knowing where the center of gravity lands on the ground is half the battle; the other half is finding a reliable transit point for it in the upper part of the body. No need for extensive new research, however, the Greek sculptors discovered where to locate the center of gravity in the upper body over twenty-three hundred years ago. In the Italian Renaissance it was rediscovered and has been passed down sculptor to sculptor ever since. My teacher, Rapheal Sabatini, "Boss" we called him, taught me where to find it, and now, I'll show you.

Stand with you feet apart and on line with each other. Put a finger at the base of your throat where the collar bones come together. Point your finger directly towards yourself in an accusatory gesture, and look down towards you feet. Now shift your weight back and forth between your feet and you will discover that the base of your throat centers exactly above where the weight goes. **A.** (see figure on P 58)

Most of us have little trouble standing balanced on one foot for a short time. If you were to do this you would know, again by deduction, that the only place your center of gravity could be located is at the arch of this supporting foot. Again by deduction, we would know that our standing leg, which sockets into the outside of the pelvis, would need to angle rather sharply inward, under our body, so our foot can find a place directly below the center of gravity. **B.**

Statue of Hagias the athlete, marble copy of a bronze statue made by Lysippos in 340 BC. Delphi Archaeological Museum.

This idea that the support leg angles under the body was a hard one to sell to my figure modeling students. They, like everyone else, including researchers think of the legs as coming straight down under the body. This is only one of a number of embedded image fallacies that all humans accept as fact. This fallacy shows up in the concepts and images within the literature of physical anthropology whenever issues of human balance and walking arise .

As you balance on one foot you will discover that your muscles are not working as hard as you would have thought to accomplish such a feat. As we stand normally, we habitually take the weight on one leg, locking up the joints of that leg so the weight is conveyed by bone and joint rather than by muscles. Standing is very energy efficient. M.M. Abitbol, writing in the American Journal of Physical Anthropology No. 77, 1988 found that it requires only seven percent more energy for us to stand than to lie on our backs.

In our hang-on-the-corner mode we stand on one locked leg using the other as a balance prop. We never think about it but when the pressure starts to build in the joints of the standing leg we switch to the other. The prop leg is bent at the knee and has its foot flat on the ground. To achieve this it is necessary to drop one side of the pelvis. This brings us to the second part of the sculptor's inquiry: the parts that can't move.

It is relatively easy to see what can move and what is immobile in the outer extremities but within the torso it is a bit more difficult. There are three relatively immobile blocks in the body. The head is one of these blocks with little of no internal movement except for the workings of the jaw. The head is connected to the body by a flexible shaft, one that both bends in three dimensions and swivels on an internal axis. This flexible shaft is made up of the seven cervical vertebrae that connect the head to the chest cavity.

The rib cage is another block. When the spine enters the rib cage it becomes largely immobile. The thoracic vertebrae articulate with the ribs and are knit to them with ligaments. The ribs, in turn, are locked together in the front, at the sternum. Because of this double locking front and back, there is very little movement within the rib cage.

Below the rib cage there is another zone of movement: the five lumbar vertebrae which connect the rib cage to the pelvis. The pelvis is the third block. The pattern in the upper body is therefore: block--flexible shaft--block--flexible shaft--block. The proportion of the cushioning disc material between the vertebrae in the various sections of the spine closely follow this description. There is proportionally more disc material in the neck and lower back and less in the relatively immobile upper back. When we stand and move, the flexible areas of the spine are used to shift the blocks so that the whole stack is balanced precisely over and around the center of gravity.

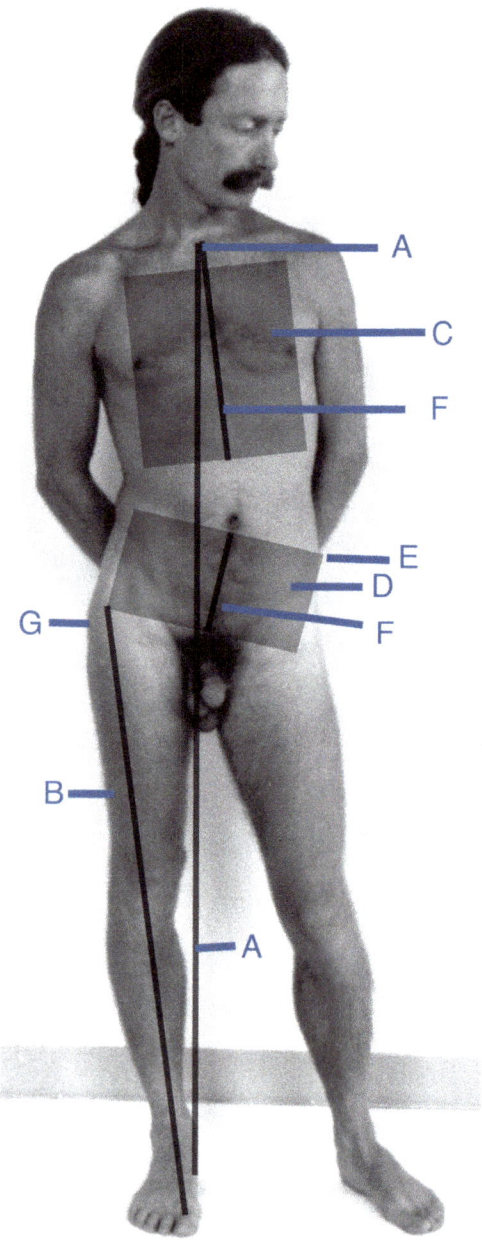

The figurative sculptor's first problem in starting a new work is to locate the pelvic and rib cage blocks. Thinking of these areas as box-like structures helps to establish the movement and the logic of the figure. The margins of the pelvic block can be found by locating projections on its surface. If you swing your hip out to the side while standing you will feel a hard bump on its outer surface. **G.** This is the greater trochanter region of the femur which is on line with the shaft of the femur and is an attachment place for major muscles. Inward from the greater trochanter, a bone neck angles upward and inward supporting the ball joint which sockets sideways into the hip. The greater trochanter can be seen on the standing leg side as it projects at the hip. This bump is the point where the line of the leg turns to become the side of the pelvis.

This turning point is used by sculptors to locate the lower margin of the pelvic block in a life model. The iliac crest is the hard bone ridge you can feel just below the waist. **E.** The crest itself, or the muscles that attach to it, can usually be clearly seen in a standing model. From the greater trochanter where the femur sockets into the pelvis to the iliac crest forms the side of the pelvic block.

A. *Center of gravity from the foot to the base of the throat.*
B. *Locked bone line of the supporting leg.*
C. *Rib cage block.*
D. *Pelvic block.*
E. *Iliac crest.*
F. *Biological center line.*
G. *Trochanter.*

The pelvic block is a logical box shape with a wide front, a narrow back, and two truncating sides connecting them. **D.** The midline of the front and the back the box are the biological center lines of the body. **F.** If we painted a center line on a cardboard box, it would move with the box. If we tipped the box, like the pelvis becomes tipped in the contrapposto pose, the center line would tip with it. In the diagraming imposed on top of the figure of the model notice that the biological center line parallels the sides of the pelvis.

The rib cage, when taken together with its overlying muscles and the suspension system for the upper limbs makes up another box-like solid. **C.** In one of the many oppositions that keeps the figure structurally interesting, this box-like solid is the opposite shape of the pelvis: narrow in the front and wide in the back. Continuing with the set of oppositions that are characteristic of a standing figure, the pelvic block tilts forward while the rib cage tilts back. This is countered by the direction and plane of the neck and the head which moves forward. These units are attached by the flexible shaft of the spine in the lumbar and cervical region. As one block moves the other two will counter move.

The collar bones meet at the top and center of the front of the rib cage. In a standing figure, this spot on the rib cage block is a transit point for the vertical axis of the center of gravity. This relationship, discovered by figurative sculptors, places this point on the rib cage block directly above the projected center of gravity's location on the ground. **A.**

SHORT LEGS

When we were toddlers, our legs were short, which we discovered by tactile self exploration. As we matured our legs grew longer both in size and in proportion to the rest of our body. Despite these physical changes, the conceptual image of ourselves and our embedded visual symbol for a human being remained the toddler's image. Unless these ideas are challenged by training, they follow us into adulthood. Seeing the figure objectively is never an easy task. When it comes to the legs, this task is further complicated by optical clues which support our preconception that legs are shorter than factual reality. Without clear and repeated instructions to the contrary, my drawing and figure modeling students usually would make the legs too short.

Classical Greek sculptors, carving marble, needed a consistent rule to assure that they would get the proportions right from the start. The Greeks were first to define these

60

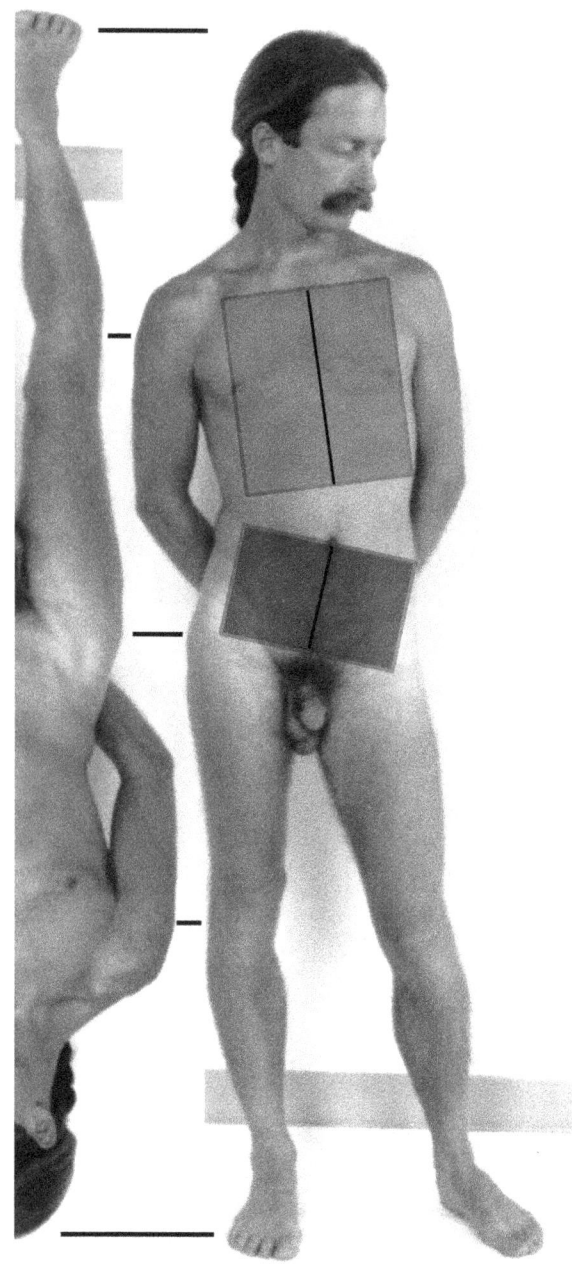

proportions, followed by the Italians during the Renaissance. My teacher, "Boss" Sabatini, coached me, and I coached my own students.

How do sculptors proportion the legs? We look on the side of the hip above the standing leg where the greater trochanter pushes out. Here we find a turning point where the directional line of the leg turns to become the directional line of the hip. This turning point is the top of the leg and halfway between the floor and the top of the head. Sculptors traditionally use a four-part proportion for the figure. The halfway point is the greater trochanter. The leg proportion is halved again just below the knee and the upper proportion halved at the armpit. My experience teaching the figure over the last thirty years has shown this set of proportions to be very consistent in a substantial number of models, both male and female. There have been exceptions but they were few and far between.

I suppose that the relationships that classical artists used to describe the figure appear subjective. The intent of classical sculptors, however, was to achieve beauty through truth. Here, for a moment, we need to disconnect the concept of truth from that of realism. It was the Romans who were most responsible for connecting truth to realism; the early Greeks never quite linked the two. To them, the human figure was used to represent the Gods. The Greeks expected to find, just like a Greek temple, a logical continuity in the relationships within the figure. The classical sculptors wanted to know what was true of all figures: how the human figure balanced in space, its proportions, its movement, the limits of its movement, and the relationship of muscles to movement. They learned that there were consistent proportions including how the leg's proportion relates in to the body. They looked for underlying unifying principles.

In many ways, these ancient Greek artists made a much more fruitful study of the figure than has been the case in the modern discipline of physical anthropology. While the ancient study worked from the outside in, and dealt comprehensively with the figure as an organic whole, the modern study works from the inside out and attempts to build an understanding of the whole out of a myriad of details. The ancient study looked for unifying principals. The modern scientific study is more interested in the differences.

Anthropologists, using Natural Selection, expect to find a continuous line of changes from ape to man. Differences in bone forms are often linked in a structure of linear evolution without further study of how these shifts in bone form fit the physical needs and habits of living primates. In the case of the heads in this study, I have focused on how changes in the physical orientation of primates have resulted in structural form changes. In the scientific study, even when bone structures are extrapolated into ideas about living humans, there is seldom a second part of the process that looks at living humans to see if the extrapolation works.

The Greek sculptors spent two hundred years using careful observation to reprogram their understanding of the human body to overcome the conceptual figure and its embedded image fallacies. Today, after centuries of modern scientific study, the image fallacies of the conceptual figure are still very much alive and raising havoc with scientific methodology and conclusions.

In studies of various ancient populations, anthropologists have a need to determine the height of individuals using only fragmentary bone evidence. They now do this by using rather convoluted mathematical formulas. Anthropologist start with the bones and extrapolate out to the living form. Wouldn't it make more sense to first study the living figure using work that artists have already accomplished? I wouldn't be too surprised if a study would reveal proportional differences between groups of people, but establishing a standard would be a significant first step. Once consistent proportions for living humans have been determined and then relating these to the underlying bone structure will make working from fragmentary evidence both easier and surer.

Some artists' formulas use the head as a unit to determine proportions. This is not as reliable a unit as leg length since that is constrained by the function of walking.
Physical anthropologists have proven to be much less evidence based in their work than

were the classical sculptors. It should be fairly easy to get a large data set of body proportions using a standardized photographing protocol especially on a large college or university campus. College students should be ideal because they are fully mature, well nourished, and racially and ethnically varied. Establishing a standard protocol would make it possible for researchers in the field to compare local populations against the student data to determine if local conditions are a factor in body proportions.

WALKING THE WALK

One of the most important aspects of human evolution is bipedal walking and it would seem to be the one most accessible to rational study. We all walk and further we have had access to solid photographic imagery since the nineteenth century. When Eadweard Muybridge and the American painter Thomas Eakins first did stop-action photography of people walking.   Eakins, being comfortable around nude models and using one who was well seasoned, achieved images that would still be useful today. It is surprising, with such easy access to factual imagery, the presentation of walking in the literature of physical anthropology is still carrying conclusions based on figurative image fallacies.

We walk using the same set of mechanisms and balance structures that we use to stand. Just as when we stand, we have to assure that the vertical load is balanced over the center of gravity. When we walk, we lock the joints of whichever leg is taking the weight. This minimizes the use of muscles to support the vertical load, and makes bipedal walking very energy efficient. Human walking is more energy efficient than the quadrupedal mode in animals. This efficiency gave us bipedal walkers an advantage. In our evolutionary history, the advanced foot appeared before the advanced brain.

In the walking figure, the pelvis and rib cage are held relatively level because the stepping foot is raised off the ground. The major counter movement in walking is the arms swinging opposite to the stride.

The motion of the striding figure is like a segmented wheel. The walking leg locks at the knee and rotates at the ball joint of the hip becoming like a spoke of a wheel. The foot functions as a segment of the rim of the wheel. As in a wheel, the rim segments are in line and, just like a unicycle, they are directly under the center of gravity.

In the graphic on page 60, I flipped the standing leg to demonstrate a well worn artists' proportion. The distance from the greater trochanter at the top of the femur is half the height of the figure. The hub of the wheel on the walking figures on the following page is centred on the greater trochanter. Not only are we a wheel, we are a big wheel.

A toddler, like the schematic figure, has a block like torso with little or no compensating movement between the pelvis and rib cage that can be used to achieve balance. The toddler must swing the upper body over the foot that is taking the weight and then back to the other side on the next step. The young child, in other words, must tip or toddle to walk.

In the section on bipedal locomotion in <u>An Introduction to Human Evolutionary Anatomy</u>, Leslie Aiello and Christopher Dean's extensive text, there are two illustrations on page 269: 14-23, 14-24. Below are these two drawings from other earlier studies, one after Carlsoo and the other after Sanders. Both based on the expectation that the feet are spread laterally apart when we walk. If Carlsoo, Sanders, or Dean and Aiello, for that matter, had taken just a moment to walk like the illustrations they would have quickly learned that something was wrong.

If you try to walk using either of the stride patterns you will quickly discover that you have an excessive wobble movement in the upper part of you body. This is another case where the schematic human figure has subverted scientific understanding.

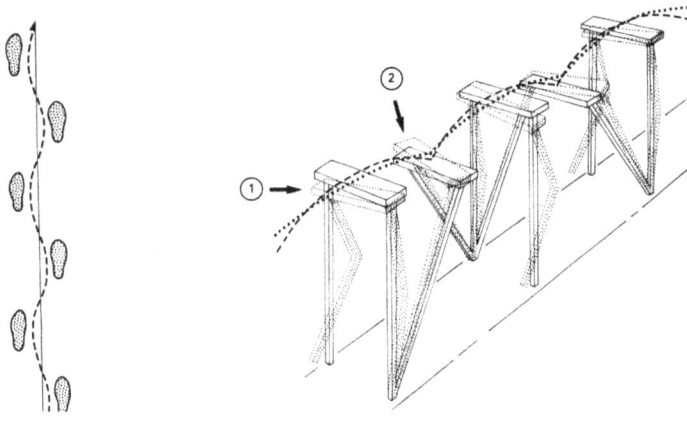

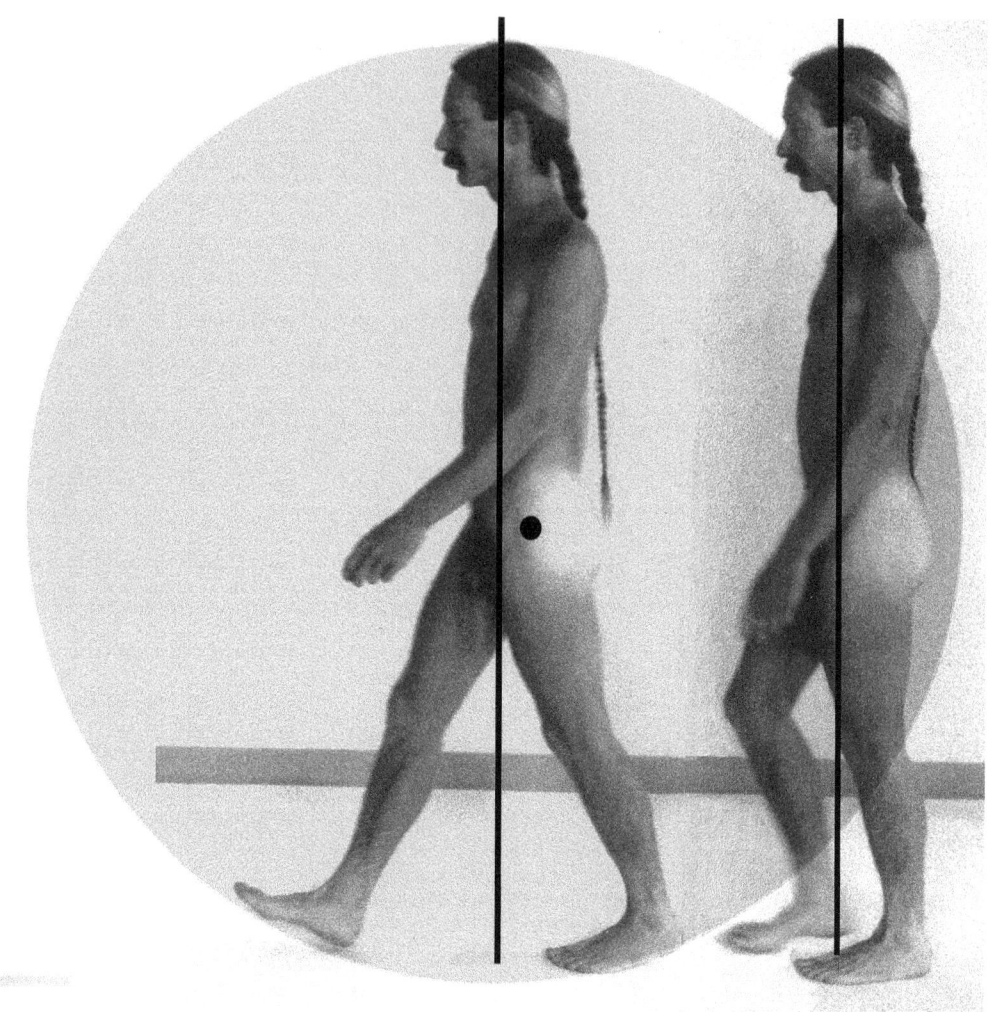

These two illustrations demonstrate the efficiency of walking. The model walking forward brings the inside of the feet into line. Go to the beach early in the morning and the fresh foot prints you will find there line up. Some folks will point their toes outward to get better stability but their heel imprints will be in line. If the foot prints are like the illustration on the previous page it is a good bet that person was heavy or old or both.

Sculptors read turning points the apex where form turns and where the light breaks. Notice where the light breaks on the forward leg moving from the hip inward to the knee then in the lower leg curving in towards the center of gravity at the ankle. This is the power line and it continues through the foot to the big toe. This is easier to see on the back leg.

BASEBALL

My father would have loved it if I had played baseball. He was an avid fan and student of the game. I, however, was a large kid and too slow for baseball, so when I started junior high school I became a tackle on the football team and threw the shot put in spring track. By the time I got to college, I was small for these positions and sports, but I was still slow so I stayed at tackle and the shot put.

Shot puts are essentially cannon balls which soldiers and others in the nineteenth century would chuck, probably competing for beers. Being heavy, the shot is nestled in the crux of the neck and shoulder and is pushed out from there, hence the term "put". Going up through the ranks, I graduated from an 8 pound shot in junior high school, to a 12 pound one in high school, and finally to a 16 pounds shot put in college.

Throwing the shot takes 2 seconds or less and a football play from the line not much longer. My athletic career came down to a few carefully coached moves which were enough to keep me competitive against larger guys. As I thought about it later, it was pretty much a zen activity.

I could tell the quality of the put the second it left my hand. If it departed with little effort, just a parting push from my three middle fingers, then I knew that the throw was probably going to be good and the shot would float out and away. If, however, I had to push hard against it, that was guaranteed to be a mediocre effort. Good throws take almost all their energy from the legs and torso. The back leg drives, the torso rotates snapping the chest forward, and the shot comes out like a whip being cracked.

Watch a pitcher throw and you will see the same dynamic. Notice how pitchers will fuss with the contours of the mound and groom the rubber between pitches in a prayer to sure accurate footing. All serious throws get their power from the legs and the torso. Batters also harness the power of their legs and bodies when swinging.

It seems to me that pitching and batting are the essential human skills and may be one important reason for being bipedal. The two essential questions are: how did early humans survive without the substantial natural weapons of tooth and claw and why was being bipedal an advantage?

There are a number of advantages for being bipedal. It is more energy efficient and allows for carrying things, like food. Meals don't have to be consumed where they are found, allowing for the collective nurturing of a community. Lions, for instance, feed on their prey where they kill it.

Bushmen in the Kalahari in Africa were known to trot after an antelope and keep it moving until it dropped from exhaustion, sometimes 24 hours later. The antelope can run faster but the bushman can run longer. Of course, it is a group effort with individuals taking turns running the antelope in circles. Once they had it, they could carry it back to camp.

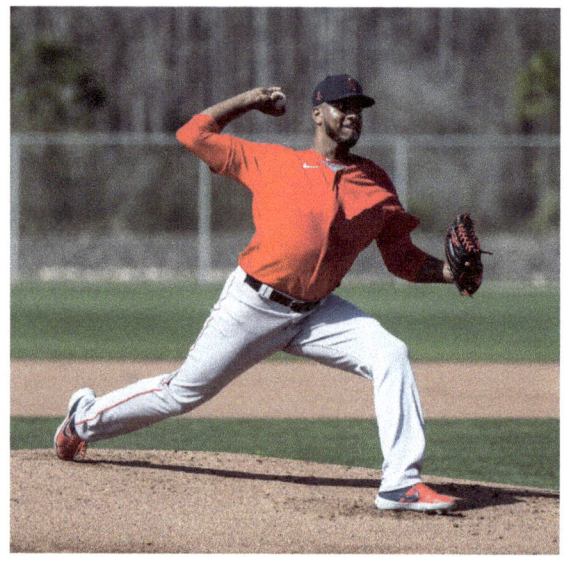

Baseball pitcher and Hong Kong protester using the same dynamics to throw.
Alamy stock photos

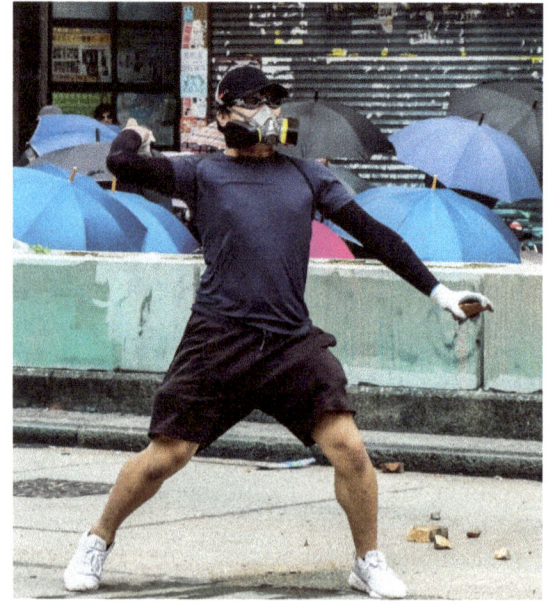

A major league pitcher can deliver a ninety mile-an-hour pitch to a few square inch target. If he hits the batter, it can cause serious injury and, if thought to be on purpose, spark a fight. A baseball weighs just 5 ounces and is smooth and somewhat resilient. A half pound of rock with some sharp edges would be very dangerous even if the velocity was slower.

If you would like to see the essential primitive baseball game, go to a riot. Riots break down into two sides: pitchers and batters, protesters and police. Protestors throw things: rocks, molotov cocktails, tear gas canisters and the police equipped with shields, helmets, and face masks move in, trying to get close enough to bat them.

If you asked the participants, on either side, if crazed bipeds throwing rocks at them or trying to beat them with sticks is intimidating, I am sure they would all agree that it is. "Sticks and stones will break my bones" as the nursery rhyme goes. I am also sure a large predator confronted by a howling mob of our ancestors throwing stones and swinging sticks would have been intimidated.

The skill of throwing with accuracy or delivering an accurate, forceful blow with a stick would also have been used in hunting. We wouldn't find evidence of this today in its earliest forms because none of the implements were modified for the task. I suspect, however, when we find an ancient hominid with a foot made for walking we likely have also discovered a pitcher and a batter.

THE STUDY

This set of study comparisons primarily uses David Brill's photos of skulls From Lucy to Language. Brill's stunning photographs provide a wealth of information. The standard position for the photos was the Frankfurt Horizontal which, while better than the Table Fallacy, still distorts the relationships between skulls. The distortion is greater when the facial stalk angle is more open to the base of the brain case as it is when there is a more horizontally orienting necks and spines. The skulls in this study have been rotated to their going-forward-alertly position.

Imagery is being used like statistics and require a base element to be made equal in order to compare the variability with other elements. In most instances, the brain case was used as the base element. In each primary comparison, the overlying skull was scaled so the brain case was the same size as the comparison's brain case and oriented to fit the outlines as closely as possible. This provided a shape comparison of the two brain cases. This scaled image is compared with other elements in the two skulls to show their relative proportions. In each comparison, the top image is made somewhat transparent so the elements of the underlying skull can be seen through it.

The brain case comparisons across the entire group, including the Neanderthal Shanidar skull, reveal that the front and upper portions of the brain case, the frontal and parietal bone elements, are very consistent in shape. This consistency likely results from this portion having the unchanging function of being a container for brains and not having any kinetic function. The rear and base of the brain case is more variable being an attachment place for the nuchal muscles containing the foreman magnum and the occipital condyles.

Once the brain cases are justified, many other elements are then shown to have consistent proportions and design. With the static elements identified, it becomes possible to determine where modifications happen. Finding the correct proportions is the goal. Two skulls may be a different size and still be in proportion to each other.

The greatest variant in facial stalk angle is in the Inuit and Zaire (Congolese) skulls. The facial stalk angle normally opens a bit during the human growth cycle as the body and head proportions change. The flexibility of this relationship in primates is best demonstrated by gorillas which reposition the angle of their facial stalk during maturation (see The Gorilla Growth Series). Surprisingly, as the comparisons reveal, eye sockets do not have a fixed position in the brain case but float up and down as the growth or absorption of the frontal bone drives the rotation of the facial stalk.

Modern skulls are also justified with Shanidar's by overlaying the outline of the eye sockets. In most of the modern skulls, the eye socket contour fits Shanidar's almost perfectly. The eye configuration is stable, being held in tension between two functions. The outside edge of the eye socket limits the field of vision to the rear. Locating it farther rearward would open up that sight line but being a conduit for mastication forces, holds it forward. The tension between these competing functions keeps the configuration of the eye socket static across different hominids and even other primates.

Justifying the eye sockets brings the ear openings into alignment, lines up the first molars, and makes the bite plane of the teeth parallel but not necessarily at the same level. There is a great consistency in the size and shape of the zygomatic complex in all the skulls including Shanidar's. This consistency results from these elements being part of the buttress system for the molars. Eyebrow ridges are also part of the support system but they are only required when the forces are projected into the space in front of the brain case. In the comparisons with Shanidar, notice that the skull caps of the modern skulls always cover the eyebrow ridges and the space above them.

The eye socket, the zygomatic complex, the molar region and bite plane orientation along with the opening of the nasal cavity are linked elements of the upper portion of the face. The only areas of the facial stalk where Shanidar is out of proportion with the modern skulls is the extraordinary thickness of the upper dental arch and thereby in the length of the face. Another comparison justifies the molars and bite plane. It is very interesting how much the teeth compared in size and position. Once the teeth are justified the forward edge of the ascending ramus also becomes justified and the post optical bar of the two skulls line up.

A constant attribute of primates is that the zygomatic process points to the first molar. The Neanderthal has pulled the rear molars forward. Neanderthals have pushed this arrangement forward so the point of the Zygomatic process points a bit in front of the first molar.

The differences between Neanderthal skulls and those of the present human population is the result of occupational stresses, the Neanderthal's not our's. Except for the facial modifications caused by stress, Neanderthals were fully human.

SHANIDAR NEANDERTHAL

The composites with Shanidar and modern skulls reveal a number of things. The Shanidar brain case is a bit longer than the brain cases in the modern group but not nearly as much as expected. The modern brain cases are a very good shape match for Shanidar's. Except for the slight bulge in the occipital region, it is quite similar to the modern ones. The variability of shape within the group of modern brain cases is as great as the variability they, in turn, have with Shanidar. The angle of the facial stalk to the brain case is also within modern parameters, being well inside the extremes of the modern group.

Shanidar's facial structure is set out in front and almost disassociated from the rest of the skull. There is an extraordinary robustness of the upper dental arch which is much thicker than those in modern skulls. The depth of the maxillary accounts for almost all of the extraordinary length of Shanidar's face.

The Shanidar jaw is longer than all the modern ones, unusually so, considering that jaw length in primates, is primarily a factor of the angle of the facial stalk to the brain case. This, in turn, is a factor of posture. Horizontally orienting primates have long jaws and those with more vertical postures, shorter ones. The wider the angle, the longer the jaw. Shanidar's facial angle is more open than the Chinese and Inuit ones, more compressed than the Zaire and German ones, and closest to the Bengali and Solomon Island skulls. Shanidar's jaw is, however, longer than any in the modern group and longer than it would normally be because the face has moved outward, away from the brain case.

The unusual configuration of Neanderthals' faces results from the consistent use of their teeth, particularly the front teeth. This use of the mouth has clear ramifications in the structure of the face and in the shape of the skull. The buttresses for the canines in the front of the mouth are along the nasal cavity. The compressive force lines from the canines can be clearly seen in chimpanzees and gorillas which have very large canines. In humans, these support structures are also observable but are less obvious. In modern humans, the two central incisors direct force towards the base of the nose and are contained by the upper dental arch.

Neanderthals use of their mouths for work caused the restructuring of their face to buttress these forces. The extraordinary thickness of the upper dental arch is part of the strengthening of the face. Neanderthals flattened the front of the dental arch shifting the force lines which may account for the puffy quality of the face noted in the literature. Also noted, is the projection of the mid-face which functions as a truss form for the front of the mouth.

The lower jaw is subservient to the upper dental arch. If the facial stalk swings outward, the jaw must follow and grow longer. Jaws are third class levers so the longer they become, the more mechanically inefficient they are. Generally, in primates, the incisors are cutters and perform a nipping function. Normally when jaws grow longer, the incisors splay outward in order to place the mouth in its proscribed place with as little investment in structure as possible. In the Gorilla Growth Series the incisors splay out as the jaw grows longer during the growth cycle. This effect can also be seen to a minor degree in the Brill skull from Zaire and a bit in the Bengali skull. Because of the visual environment in which they live, West Africans tend to have a more open facial angle and therefore a bit longer jaw. In the Zaire jaw, which is still not as long as the Shanidar one, the front teeth splay out while Shanidar's are quite vertical. Modern humans use just the four incisors to cut but Neanderthals had a flattened jaw and used both canines and, even to a large extent the front premolars to do the work in the front of the mouth. Whatever they were doing, they needed 16 teeth to be part of the action instead of just eight.

It appears that for their hundred and seventy thousand year history in Europe, Neanderthals were constantly taking something in their mouths, clamping down hard, and then pulling backward on the object. We can tell that the motion was primarily backward because of the occipital bun on the back of Neanderthal's brain case. The muscles that attach to the occipital region are the ones that raise the head and would, in such an action, be the primary actors. Neanderthals, with large brain cases could provide additional surface area for muscle attachment by deforming the rear of the brain case rather than raising a crest which is the case in some primates. Neanderthals show biomechanical consistency in their facial structure, wear patterns on the teeth, the configuration of the jaw, and the occipital bun. If they had been using their mouths to twist or wrench side to side we would expect to find more robustness in the zygomatic arch which buttresses against lateral forces. The zygomatic arch in Shanidar is actually less robust than in some of the modern examples.

This heavy use of the front of the mouth by Neanderthals and Heidelbergensis is unique in primates. The front of the mouth is the region of least mechanical advantage. Primates mouths are designed to cut with the front teeth and grind more heavily with the molars. The normal force gradient is evident in the ascending surface area of teeth from front to back. Facial structure in primates, including modern humans, has the function of containing these normal primate chewing forces. Neanderthals changed the primate pattern of using the mouth and thereby their facial structure. The strong similarities between Neanderthal and

Heidelbergensis wouldn't be as evident if the occupational use of their mouths were not also similar. Form will not be inherited for long if use and occupation are not also inherited. It seems likely that Heidelbergensis and Neanderthal are related but the similarities may be simply a convergence of form.

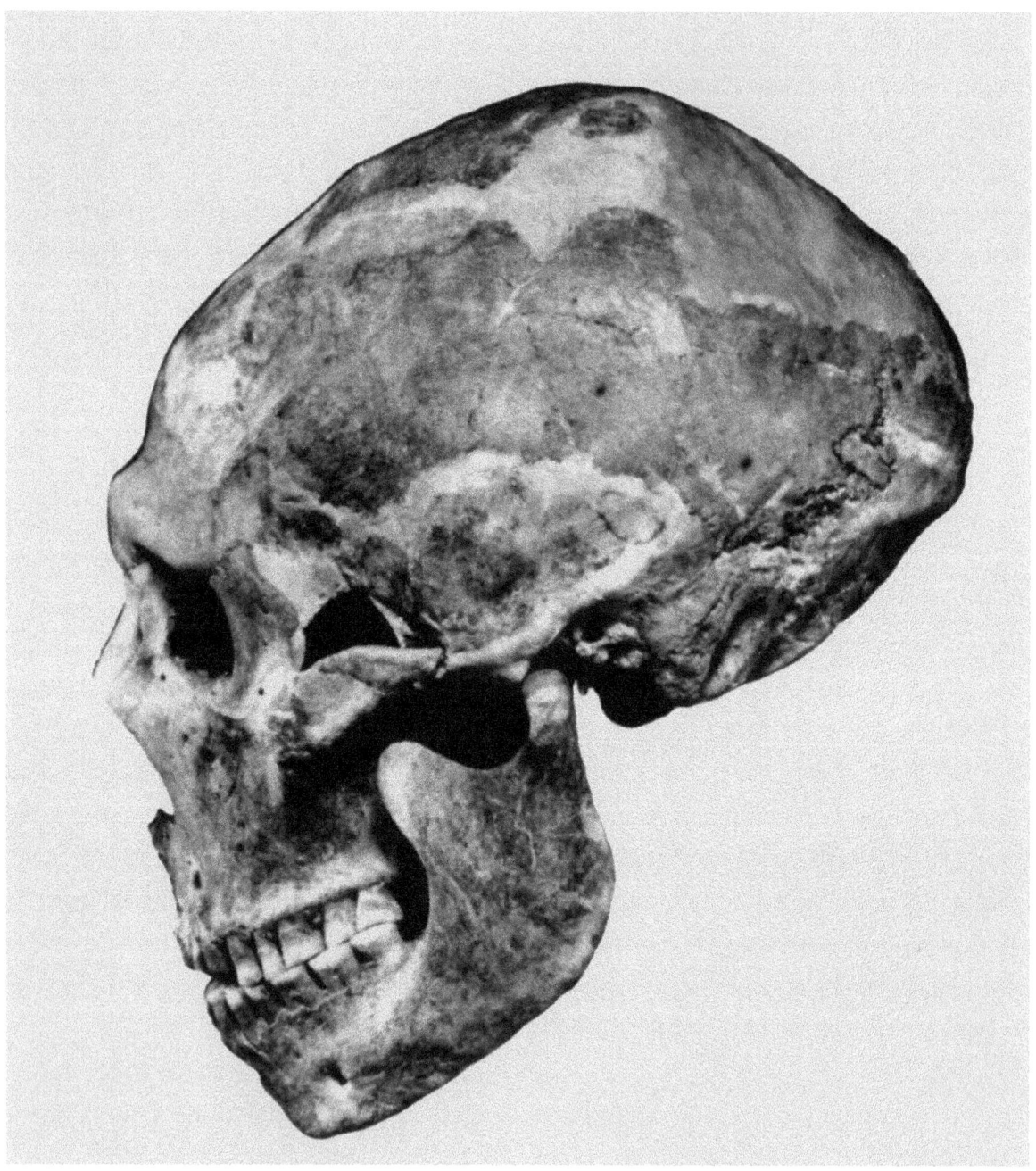

Shanidar, Neanderthal skull from what is now northern Iraq. This skull has been put into the position it would have been held in life while moving forward. Much of the descriptions of Neanderthals in the literature are the result of the Table Fallacy. They are described as having elongated brain cases, with sloping foreheads, and prognathic faces. Directly comparing Shanidar with modern skulls by overlaying and justifying images disproves all of these assertions.   photo: Donsmaps.com

The lab skull has been overlain on Shanidar Neanderthal with the brain cases justified. Comparing Shanidar brain case with other modern skulls yields similar results. Only a small bulge in the occipital region makes Shanidar's brain case longer. The architecture of the braincase is the same with the ear canals in the same position. Shanidar did constant work with the front of the mouth which is buttressed by the face. The face moved outward to provide more supporting structure.

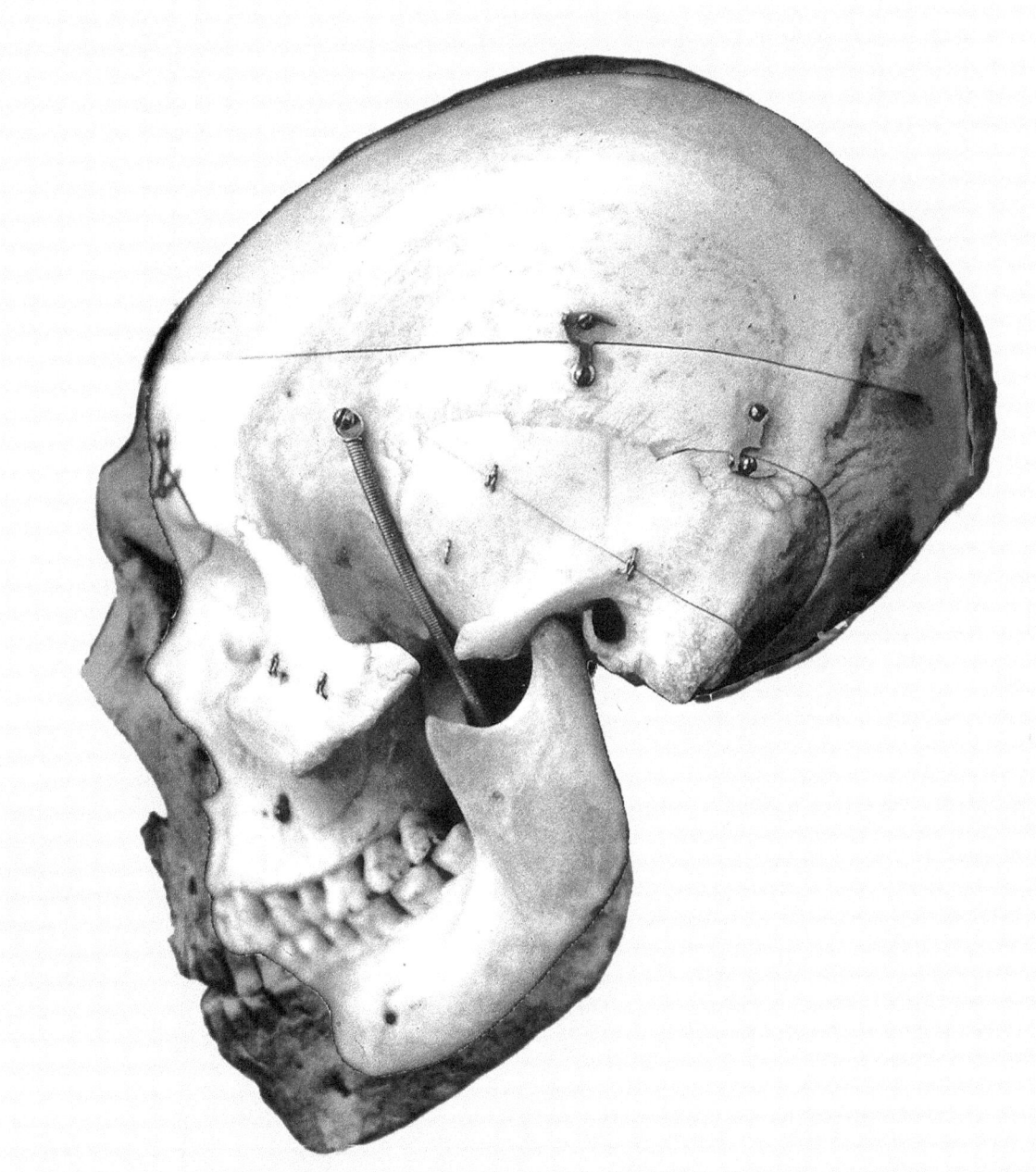

Pulling the face forward has disassociated the buttress line from the molars to the brain case producing substantial eyebrow ridges. The lab face and Shanidar's are parallel which indicates they occupied a similar visual environment. The origin of the lab skull is unknown but laboratory bones of that era usually came from India. Brill's Bengali skull has a brain case that also closely conforms to Shanidar's.

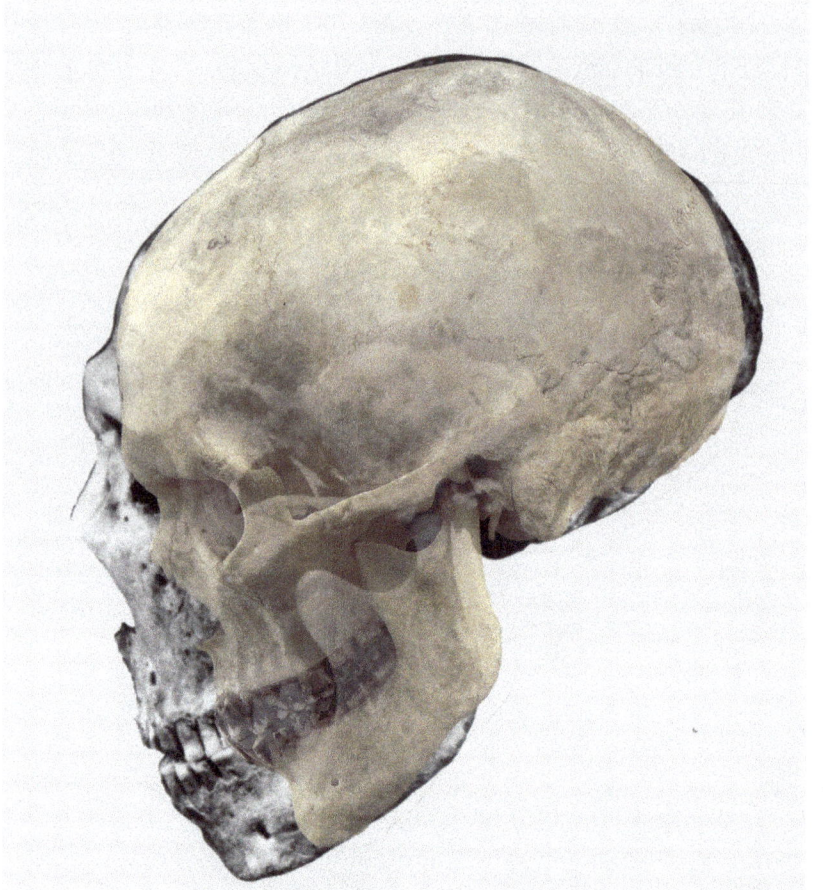

The Bengali skull is overlain on Shanidar's and, like the lab skull, is a very good match. Again there is a slight elongation of the occipital region. The Bengali's facial stalk is slightly more rotated rearward than Shanidar's which indicates this person occupied a more open visual environment.

Brill's Inuit is overlain on Shanidar with their brain cases justified. On all these brain case comparisons the bregma, the point where the sagittal and coronal sutures meet, are concurrent. Like the comparison between the Zaire and Inuit skulls, Shanidar's facial stalk has rotated outward indicating a denser visual environment. Except for a slight elongation in the rear, the outlines are a near perfect fit as it is with the Bengali skull above. Unlike other modern examples, the Inuit's ear canal and jaw fossa are compressed back in the skulls architecture. Shanidar's ear canal is under the Inuit's jaw fossa. Shanidar has a normal brain case architecture while the Inuit's is modified.

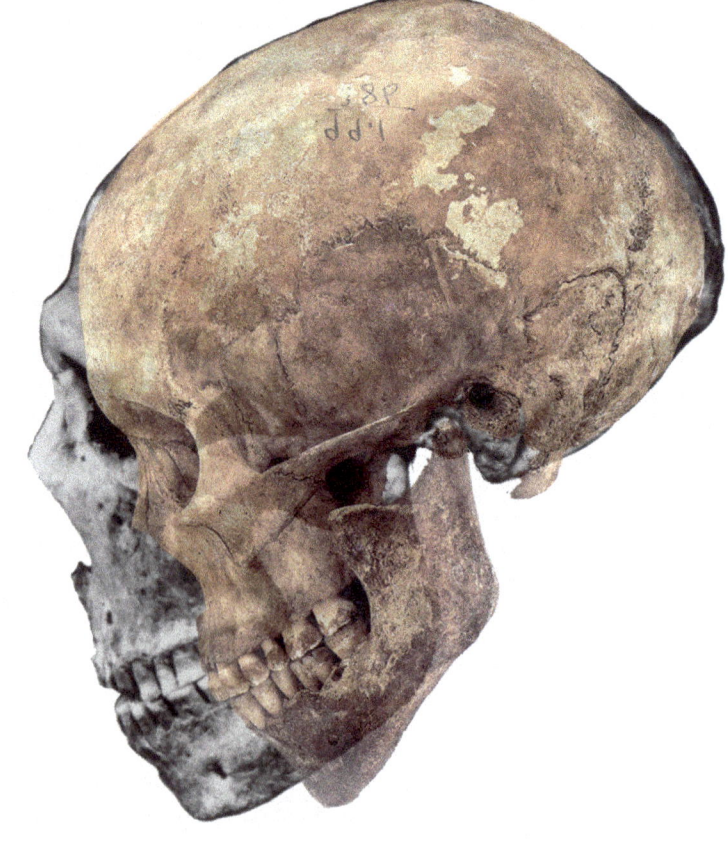

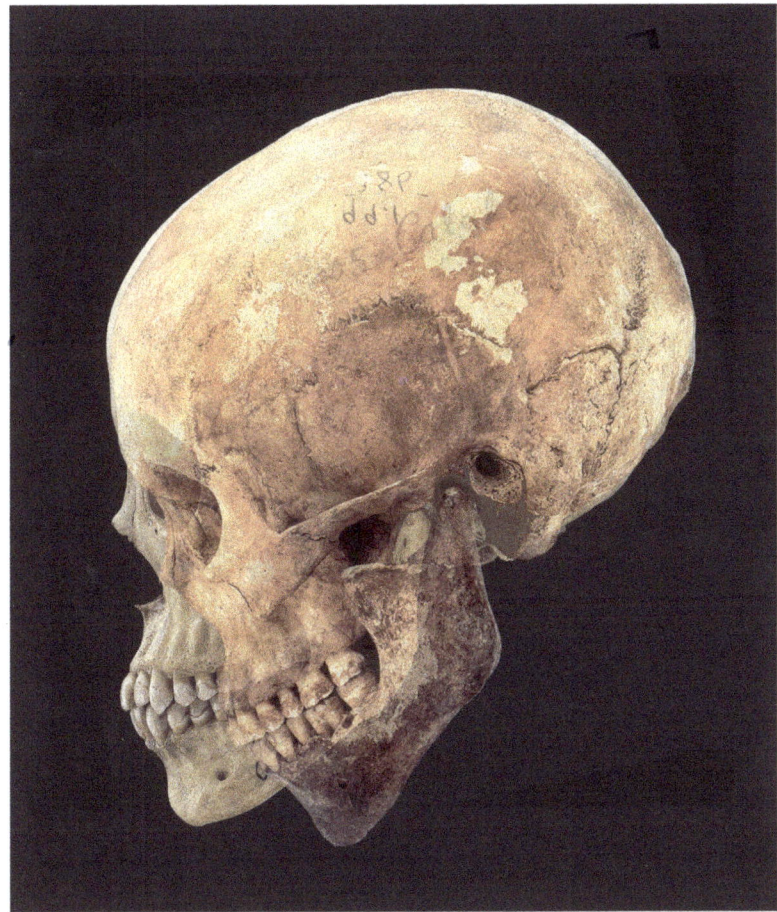

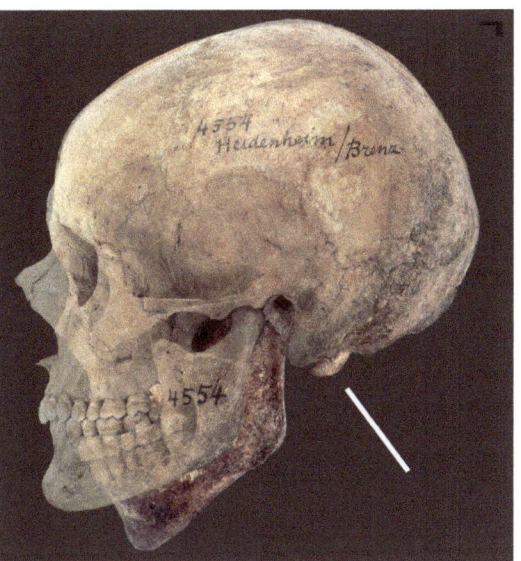

This image compares an Inuit skull with that of a Congolese. It is a repeat from earlier in the book because of it fits the context in both places.

Above is Brill's German skull overlain on the Inuit's. Like the Inuit Congolese comparison the brain cases are an almost perfect match. One point of divergence are the mastoid processes which are angled differently because the Inuit habitually held the head higher. This caused the rotation of the facial stalk and drove the ear canal back in the braincase architecture.

INUITS AND SOLOMON ISLANDERS

Much has been written about the convergence of traits between Inuits and Neanderthals especially in terms of tooth wear. Inuits have the reputation of using the anterior teeth as a tool to manipulate material. Comparing the images of Shanidar and Brill's Inuit skull reveal some interesting convergences and differences. The Inuit jaw has a unique shape being thicker in the front. When the two jaws are justified the Inuit jaw is actually more robust in the chin region but is reduced in thickness towards the molar region. Shanidar's jaw, on the other hand, is uniformly thick along its length. The Solomon Island skull's jaw is quite like Shanidar's, it is a bit smaller and equally chinless. Bone equals force. Inuits jaw shows greater force applied to the front of the jaw while Shanidar and the Solomon jaws are more uniform in the force gradient with Shanidar having applied greater continual stress to its jaw than any of the modern examples.

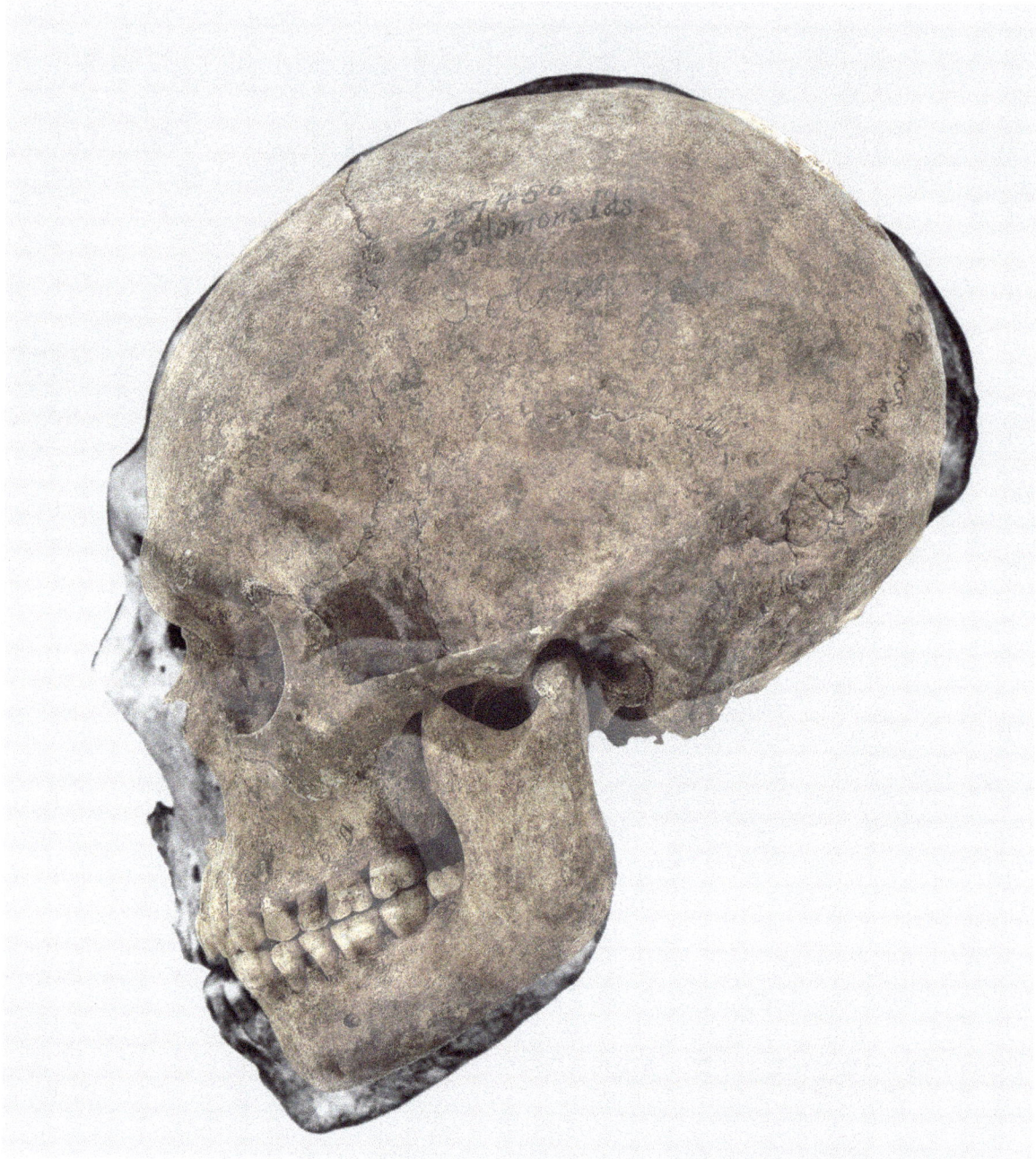

Brill's image of a skull from the Solomon Islands is overlain on Shanidar. Solomon Island is part of The Melanesian chain of islands where many of the inhabitants have between 4 to 6 percent of archaic Denisovan DNA. Denisovan DNA was first identified from a finger bone and teeth found in a cave in Siberia. Tibetans and some Inuits also have Denisovan DNA. Eurasians have 1 to 2 percent Neanderthal DNA while all Humans share 98.8 percent of our DNA with chimpanzees. I suppose the archaic DNA is part of the remaining 1.2 percent that is exclusively human. In this comparison, the two brain cases are justified and are a good match but not better than with other modern humans. With the brain cases justified, Shanidar's face is forward and longer than the Solomon skull and slightly rotated rearward indicating Shanidar living in a bit less complicated visual environment. The Neanderthal's zygomatic arch is flared due to the elongated face.

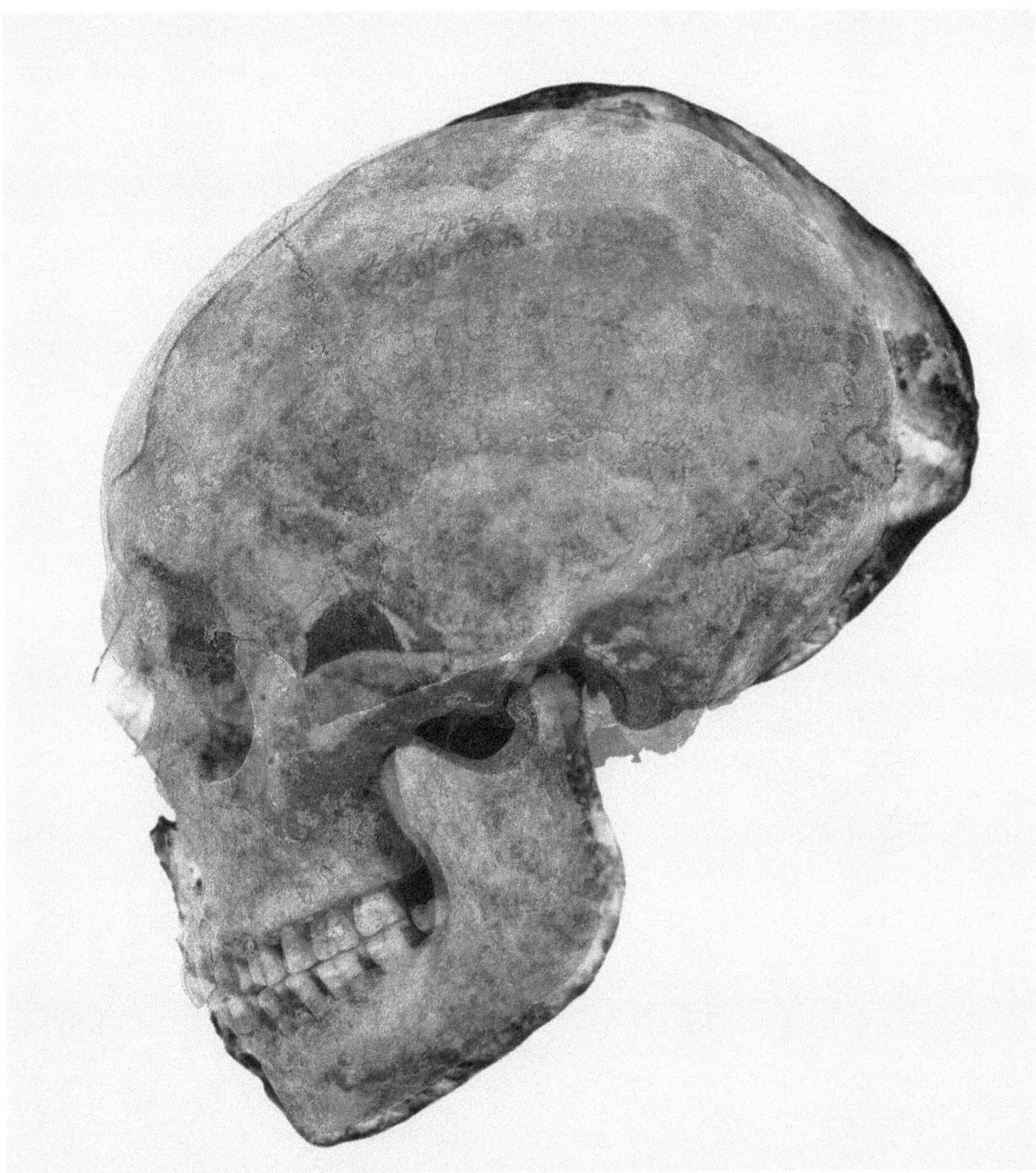

The two skulls have been justified along the bite plane of the teeth. Shanidar's face is pushed forward and unique to Shanidar, the teeth and the biomechanical buttress line for the molars are also forward. The Solomon lower jaw has a third molar showing in the retromolar space of the Neanderthal's jaw. Shanidar's tooth pattern is half of a space forward. Both post optical bar lines are lined up which confirms the biomechanical justification. The Solomon skull covers the eyebrow ridges of Shanidar thus absorbing the molar pressures and negating the need for strong eyebrow ridges. The difference between the skulls is the result of Neanderthals doing constant work with the front of their mouths. Shanidar's maxillary is twice as thick as in modern skulls which lengthens the face and pushes the eye socket higher. The face is set forward. The rear of the braincase is slightly elongated to anchor enlarge neck muscles which take part in the work being done by the mouth.

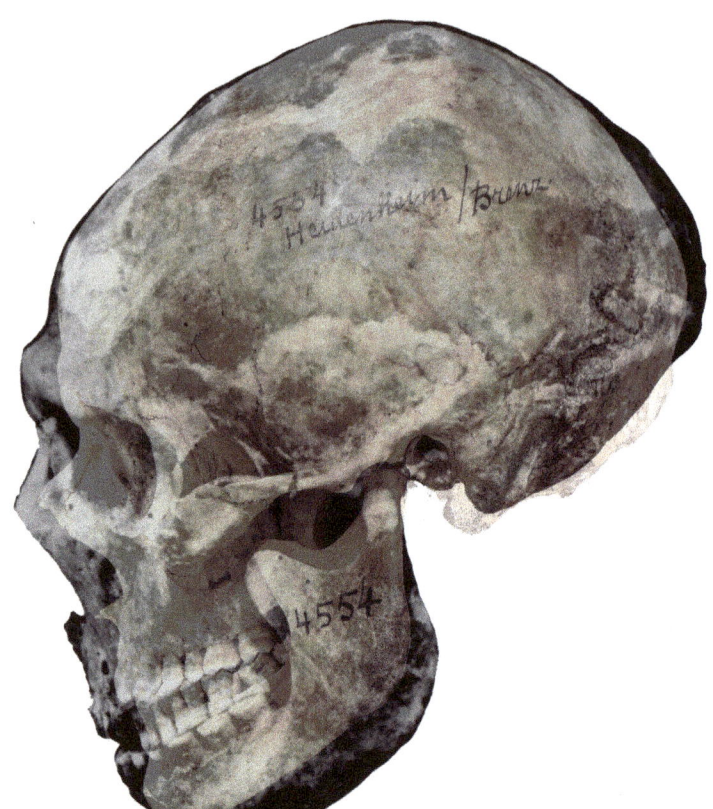

78

On this page are two comparisons between Shanidar Neanderthal and Brill's German skull. The top one compares the brain cases which are a good match for shape except for a bit larger bulge in the back of the skull when compared to other modern skulls. The back contour of the skull and the lower area towards the mastoid process diverge These are zones of kinetic activity the nuchal being an anchor for the neck muscles and the lower margin having the mastoid process and the foramen magnum, the entry for the spine. Both these contours would be influenced by the facial stalk's angle. The angle of the facial stalk indicates that the German occupied a denser visual environment than the Neanderthal.

To the right is the German skull placed on Shanidar's with their eye sockets justified. Eye socket are a very stable configuration from gorillas onward.(see page 41) The post optical bar being justified brings the bite plane of the teeth parallel. and organizes the biomechanics of the molar forces. It also brings the brain case over the eyebrow ridges of the Neanderthal demonstrating their function in absorbing dental pressure. Often, the discussion in the literature is the size of Neanderthal's nasal opening. Much has been speculated on its cause so it is surprising to find that this German skull has proportionally as large an opening as Shandar's.

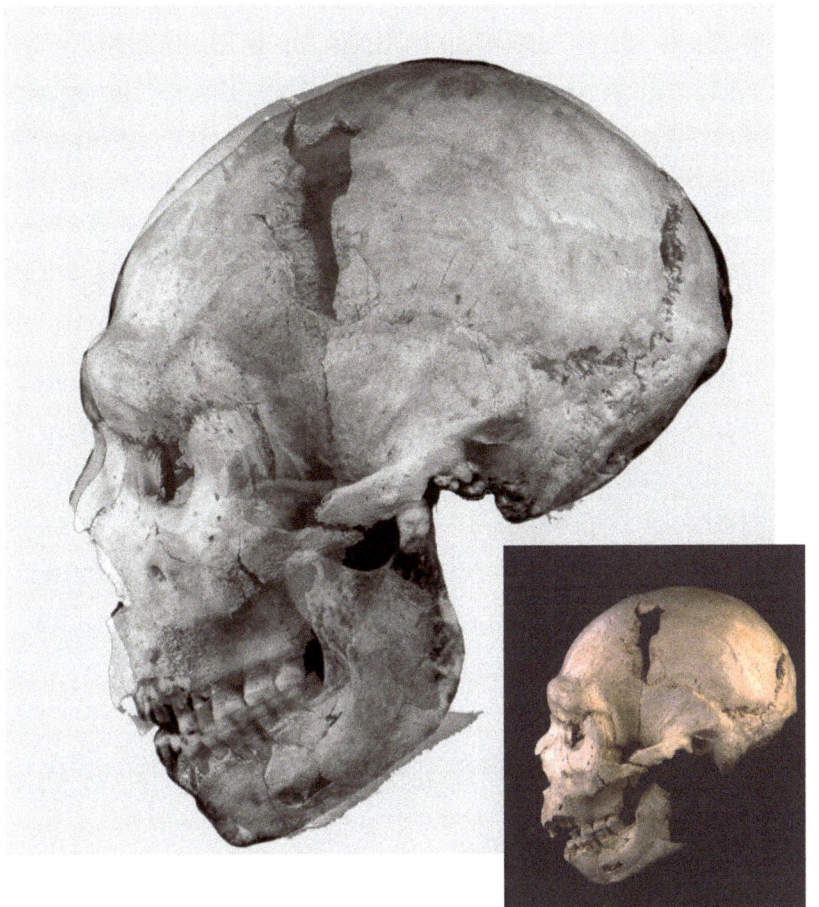

Heidelbergensis Atapeurca superimposed on the Shanidar Neanderthal skull. These two skulls are not the same size but when size justified they have a very similar configuration. The bite plane and the three upper molars are perfectly in sync in size, shape, and placement. This alignment justifies the eye sockets. The shape of the brain cases are quite similar. This comparison aligns the face but the Heidelburgensis has a bit more open angle of the facial stalk to the braincase than Shanidar.

The small Heidelburgensis skull is provided for reference.

The Inuit skull is placed on the Heidelburgensis skull with the brain cases justified. The contours of both are almost identical which demonstrates a highly conserved configuration. The brain cases of modern people also have the same profile as the Inuit skull so it is a stand in for all modern humans. Of all the studied Brill skulls the Inuit is unique in having the ear canal and the jaw's joint further back in the braincase. Heidelburgensis has the same placement as the modern skulls. The angle of the facial stalk is more acute in the Inuit which, like the comparison with the Zaire skull, indicating a denser visual environment for Heidelburgensis.

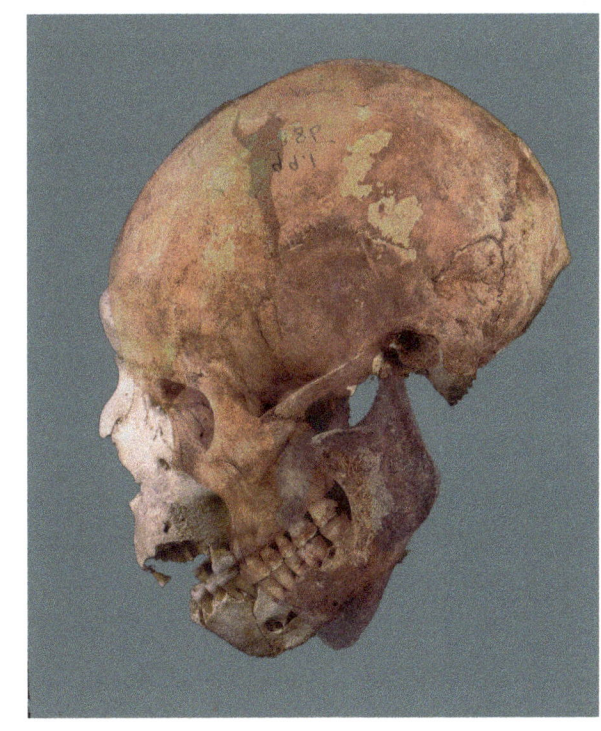

Over the years I also became interested in, and maybe a bit protective, of Neanderthals. Many researchers often quietly and scholarly denigrate of our ancestors and ancient relatives. Neanderthals have a particularly tough row to hoe because they disappeared at a time when a new population that looked like us arrived. The scholarly question often is, "what made the Neanderthals inadequate?"

Genetic scientists tell us that there were a number of populations of which no evidence exists except for tell tale bits of their DNA in people alive today. The original hunter gathers in northern Europe left a trace of their DNA in the modern population. Likewise, where I live in North Central Pennsylvania, the only remains of the indigenous population are some place names and occasional DNA inclusion in the current inhabitants. This transition took only a couple of hundred years. Like these other lost populations, Neanderthals weren't necessarily deficient but simply were shuffled off by a succeeding group with a better economic niche or better weapons. Sometimes it was war, other times it was simply being marginalized out of existence.

Searching for articles on Neanderthals, I found Academia.edu which has been a great source. Over the last year or two I have read, or at least perused, a couple hundred scholarly papers. Some were unreadable being mostly jargon but many others were clearly written for which I am grateful.

These stated attributes of Neanderthals are all attributable to heavy use of the front teeth.

Computer simulations show that Neanderthal facial morphology represents adaptation to cold and high energy demands, but not heavy biting

Philosophical Transactions of The Royal Society B Biological Sciences
Published:04 April 2018 Stephen Wroe, William C. H. Parr, Justin A. Ledogar, Jason Bourke, Samuel P. Evans, Luca Fiorenza, Stefano Benazzi,

A number of craniofacial features distinguish Neanderthals from modern humans, including a wide, tall nasal aperture, a depressed nasal floor, a wide projecting nasal bridge, a retromolar gap, 'swept back' zygomatic arches.

We will start with the depressed nasal floor. Comparing Shanidar's Neanderthal skull to a variety of modern skulls reveals that Shanidar's upper dental arch is at least twice as thick which is consistent with heavier use of the incisors. The floor of the depressed nasal cavity is, on it's flip side, the roof of the mouth which is a functional surface in mastication and swallowing. It can't move upward as the maxillary grows taller. The tall maxillary is the major ingredient in the elongation of Shanidar's face when compared to modern humans.

This succinctly states a major misunderstanding in the literature. Numerous articles have

> *Recent studies based on modern human samples have concluded that it is the shape, not the size of the nasal cavity, that primarily determines the capacity to warm and humidify inspired air [16]. It has been proposed that airway size likely relates to the energetics of the organism, whereas airway shape might be more indicative of physiology and climate [17]*

been written analysing nasal capacity in terms of breathing efficiency especially vis a vis Neanderthals. We do breath through them, but noses are primarily instruments of smell. The septum divides the incoming stream of breath into two flows and directs them to the underside of the brain case where they encounter the olfactory nerves. On the way in, the flow is warmed and moisturized by the complexly shaped concha which prepares the sample for the laboratory of the olfactory nerves. Warm, moist air, we have all experienced, enhances our sense of smell. We breath through our noses to gain sensory information but when things get serious, like when its chasing us, we breath through our mouths.

While moving forward alertly, all land vertebrates put their noses first in order to get a clean signal. Humans, with our vertical stance, are a challenge to this rule but we manage to do it by placing our faces in front of our bodies. Tarsiers, who rewrote the engineering plans for our heads, lived in trees and sight hunted insects at night. There are not coherent scent trails in trees so their sense of smell was less important. Unlike ground dwelling mammals, dogs, cats, cows, tarsiers have dry noses, just as we do. Even though our noses are four to six feet above the ground and poorly equipped at that, the ancient rule still applies and we incline our heads to put the nose out front.

PENGUINS

Penguins are the only other animal I can think of that has a truly vertical bipedal mode. Other birds and kangaroos walk with their legs under a more horizontally held body. Penguins are immersion swimmers, like seals, fish, and dolphins and have an elongated profile to glide through the water. Elaine Morgan in her 1972 book "The Decent of Woman" explored the idea that humans had an aquatic phase that, among number of other attributes shared with marine animals, gave us our elongated body.

This penguin has proportionally short legs but the stride is quite recognizably human like. They, like human toddlers, have little flexibility in their torsos so they also tod-

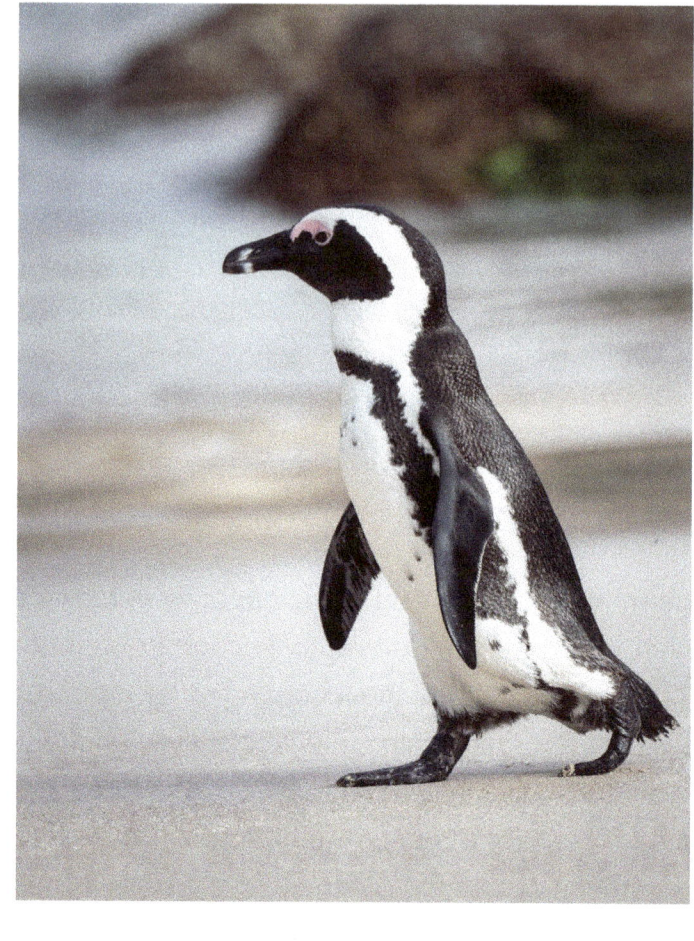

dle or sway side to side when they walk. This side motion brings their center of gravity over the stepping foot.

Penguins are both a marine and land animal. They stay in compliance with the ancient rule to lead with their nostrils and they lean forward to make that happen. Penguins hold their heads unusually high which may be a factor of modifications for swimming or be in a perpetual high-head surveillance stance due to predation from the air. Robins, which feed on exposed open ground, also use the high-head surveillance mode almost constantly.

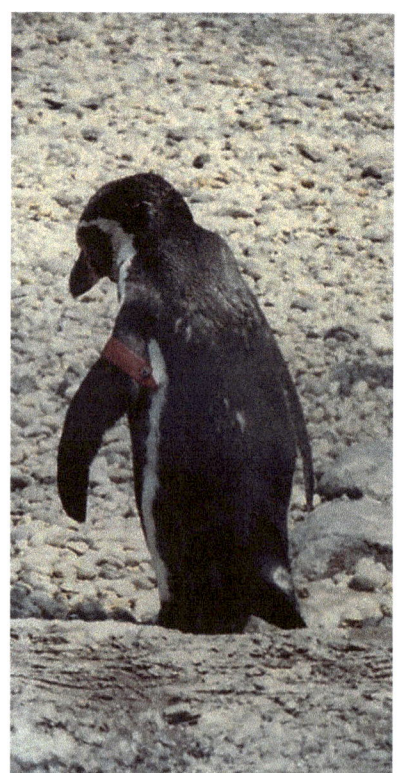

This penguin is picking her way through a stony field and drops her gaze to see where she is stepping.

PELTS

It is widely assumed that Neanderthals used animal pelts to keep warm. This appears to be a safe bet given the frigid climate and Neanderthals being a top predator. I went online to understand what preparing a usable pelt would required. Like everything else, there are videos showing how to prepare hides for those who want to do it themselves.

When it comes to bear skins, the advice from fellow hunters is to pay a professional the few hundreds of dollars to do it right. Bear hides are very difficult to prepare. The first step is to scrape the layer of fat and membrane off of the inside to the hide. The modern tool to do this is a flat blade with handles on both ends, something like a draw knife. Neanderthals likely had a bone or stone tool for this task as did Native Americans. If the membrane layer is left on, the hide will rot quickly.

After the hide is scraped, in modern times, it is rubbed with salt, a pound of salt for every pound of hide. The salt pickles the hide by drawing out moisture. Salt is a rare commodity in the wild so Neanderthals couldn't have used this step. Native Americans also didn't use salt but sometimes, when they wanted leather, soaked the hide in water mixed with wood ashes to help take the hair off.

Once the skin was scraped the "do it yourself, back to the earth folks" and Native Americans would brain the hide by rubbing mashed animal brains on it. Brains have both oils and emulsifiers which soften the hide. For the squeamish among us, a dozen egg yokes can be substituted for the brains.

Once off the animal. the fibres in the skin lock up and make it stiff and board-like. The tanner must "break" the hide by working it vigorously over a selected rock, branch, saw horse, or mounted 2X4. Experienced tanners caution new practitioners to pay particular attention to working the edges of the hide. The final step is smoking the hide over a fire which, like smoking hams, acts as a preservative.

The following excerpt from Dr. Josephine Barbe's description of how Arctic people prepared hides is likely fairly consistent with how Neanderthals prepared them. Interesting is the account of using teeth to soften the hides. Populations change but useful traditions prevail. Preparing skins was likely an existential skill for Neanderthals and may have needed to go on continuously which may accounting for the changes in Neanderthal faces, teeth, and skulls. In western Europe, the observed traits that define Neanderthals grew more pronounced over time and these traits also lessened towards the Neanderthal's end. It is interesting that Neanderthal traits are reported to be less evident in the Levant perhaps because it was warmer and therefore less need of hides.

The History of Leather Tanning
by Dr. Josephine Barbe (excerpt)
Maharam.com

One of the oldest tanning processes in existence utilizes an unusual method. People living in colder climates, such as present-day Greenland or Alaska, would work with hides, particularly seal skins, using ulo knives or special stone scrapers to remove the hair. They would then tumble or beat the skins and consequently soften them with urine. After this, the leather workers, who were primarily women, would use their teeth to masticate the hides until they became very soft and then prepare them with fat and fish oil.

Another tanning method can be reconstructed from leather garments dating from the fifth century BCE that were discovered, well preserved in ice, in Greenland. First, the hide's layer of fat was removed with clay and it was then covered with a mixture of animal brain, liver, fat, and salt. The hides were then sewn together into a round tent with needles made from bone or horn and smoked over an open fire – present in the smoke was phenol, an active tanning ingredient.

https://www.maharam.com/stories/barbe_the-history-of-leather-tanning

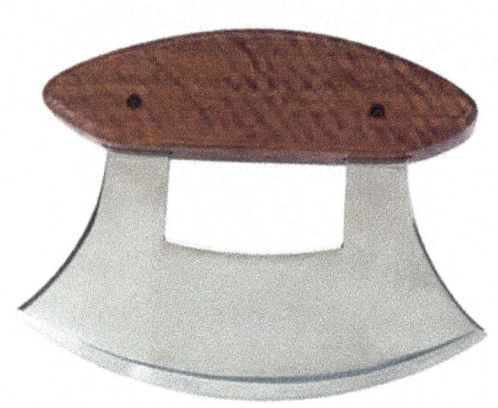

The Alaskan ULU knife was first crafted by Native Alaskan residents over 5,000 years ago and was primarily used by Eskimo women for skinning and cleaning fish.

These knives are great chopping tools for the kitchen. On the next page are stone versions with flaked sharp edges that could be used to scrape animal hides.

If Neanderthals didn't have a method to preserve the hides they used, like smoking or tanning, then a constant preparation of hides would have been necessary. It is interesting that all jaws from the Dmanisis jaws found in Georgia which date from 1.85-1.77 million years ago also have the squared off dental arch. The Dmanisis skulls also have large, robust face structure similar to Neanderthals. The African Homo erectus, Turkana Boy pages 38-39 has a much less robust facial structure. Perhaps the people of the Turkana Boy's group, being more tropical, didn't need to constantly prepare hides.

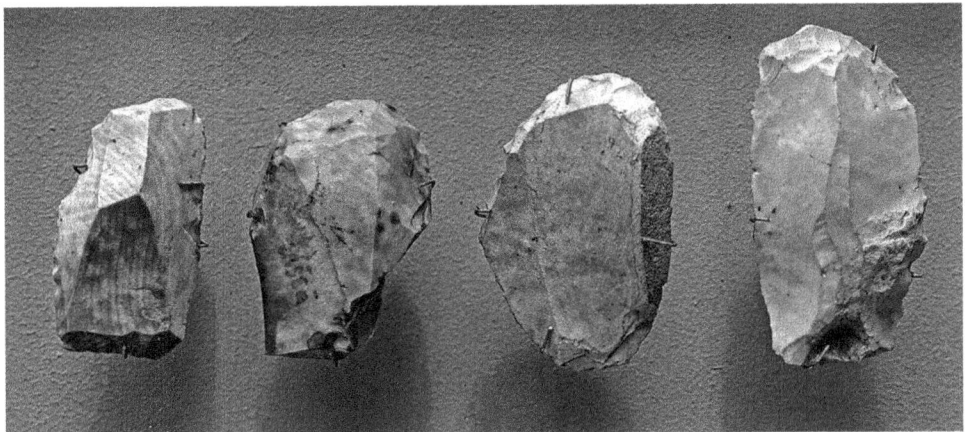

The two main tools in the Bohunician culture were the point and the scraper, four examples of which are shown here, which were used mainly for scraping skins in the leather making process, and for smoothing wood for spear shafts and other wooden tools.

Anthropos Pavilion/Moravian Museum, Brno, Czech Republic. Sourcehttps://www.dons-maps.com/bohunician.html
Author Don Hitchcock

These stone scrapers are from forty to forty five thousand years old and could have been made by either Neanderthals or *anatomically modern Homo Sapiens.*

It is easy to forget that the Stone Age lasted into the nineteenth and twentieth centuries and even persist today in the deep Amazon. There is evidence of technology swapping between Neanderthals and anatomically modern people. The question, however, is which direction the transfer went.

Given limited technical resources and the uniformity of domestic tasks over the ages, it is possible the ulu knife and the paleolithic scrapers are a convergence of form but could also be a continuity across peoples and ages.

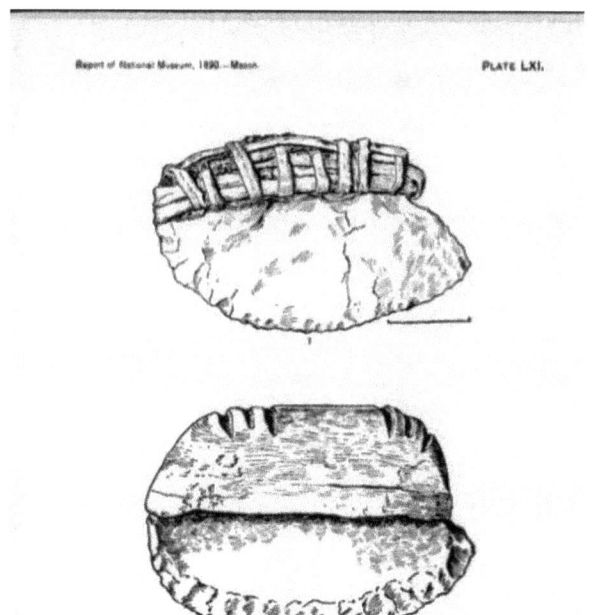

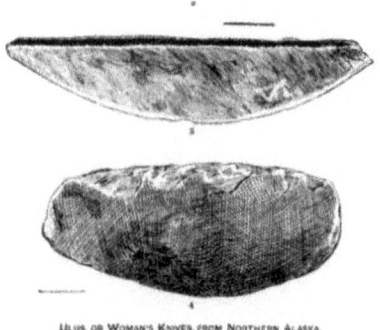

ULUS, OR WOMAN'S KNIVES, FROM NORTHERN ALASKA.

In the matter of attaching the blade to the handle or grip the Eskimo's mother wit has not deserted her. Many of the blades are tightly fitted into a socket or groove of the handle. Boas, who lived among the Cumberland Gulf Eskimo, tells us that glue is made of a mixture of seal's blood, a kind of clay, and dog's hair (Rep. Bur. Ethnol., VI, 526).

These stone knifes are from an 1892 catalog of the Smithsonian Institution by Otis T. Mason entitled <u>THE ULU, OR WOMAN'S KNIFE OF THE ESKIMO</u>. The knifes were usually made from slate but also from other stone material including a form of jade. The two on the right are similar to two of the scrapers from the Moravian Museum on the previous page.

Otis Martin reported that blades like these were found from Greenland to Kodiak Island, Alaska and could be found by the hundreds in the middens of Inuit encampments.

Photos on next page:

On top is the reconstructed Denisovan jaw which is one of just two bones carrying Denisovan DNA. The jaw dates from about 160 thousand years ago.

The Mauer jaw is attributed to Heidelburgensis and dates from two hundred thousand years ago. The jaw is also squared and the tooth wear appears to be flat.

Square Jaws

The Denisovan's front teeth below are extraordinarily worn in the front. This would be consistent with pulling upward on a hide perhaps to tighten it for scrapping or to soften it. If you put your hand on the back of your neck and raise your head you will feel the muscles there contract. These muscles attach to the nuchal area on the back of the skull where Neanderthals and presumably also Denisovans, bulged their skulls to provide more attachment surface. The comparisons of Shanidar Neanderthal with the Brill photos of contemporary people demonstrate, however, that the elongation of Neanderthal brain case is much less than thought.

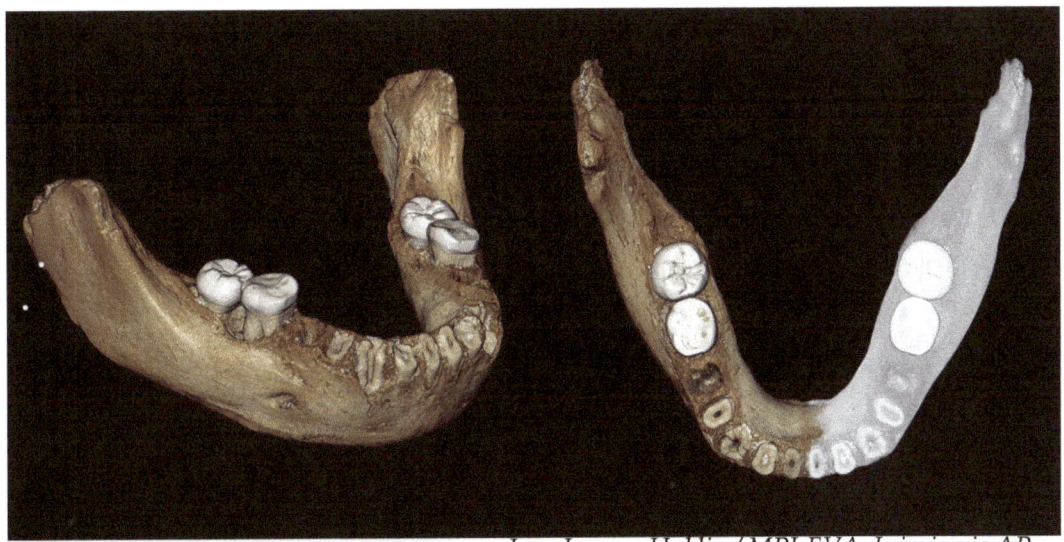

Jean-Jacques Hublin / MPI-EVA, Leipzig via AP

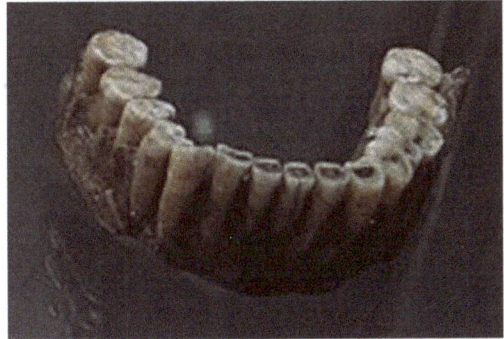

Neanderthal lower jaw from Morovia.

The Mauer jaw, to the right, is attributed to Heidelburgensis and dates from two hundred thousand years ago. The jaw is also squared in front and the tooth wear appears to be flat.

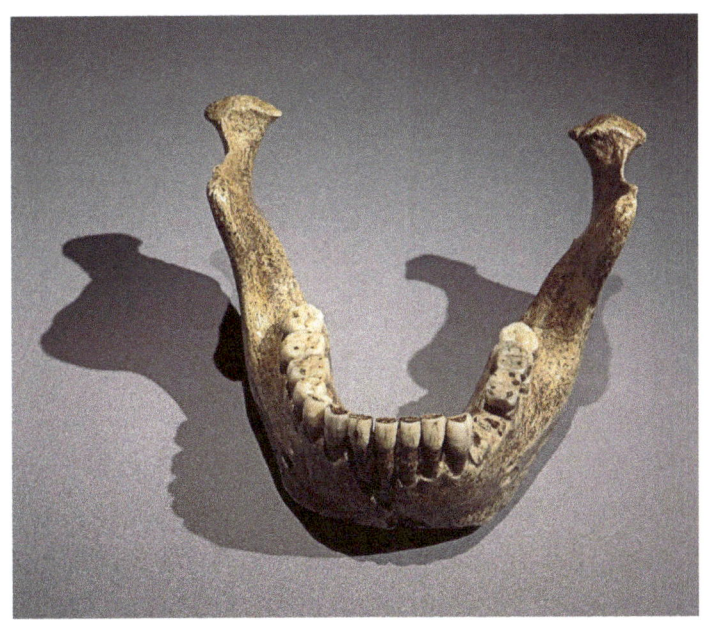

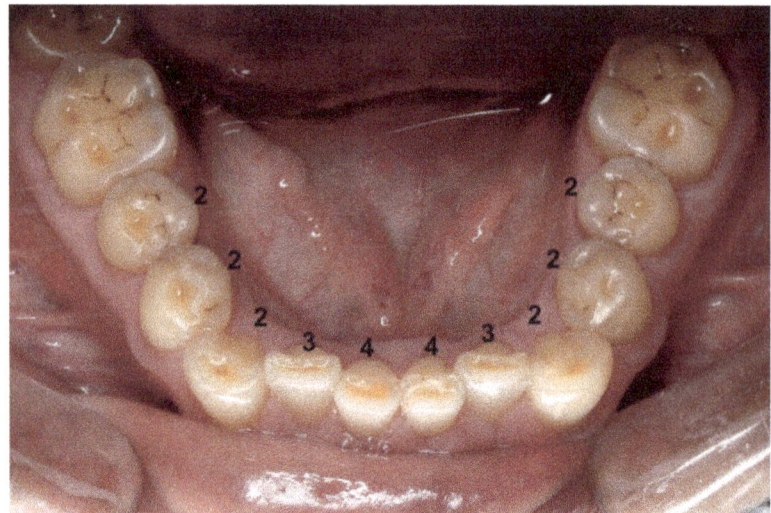

Relationship of Tooth Wear to Chronological Age among Indigenous Amazon Populations

Elma Pinto Vieira, 1 Mayara Silva Barbosa, 2 Cátia Cardoso Abdo Quintão, 3 and David Normando 4 ,*

Now here is a shocker! This is the mouth of a 48 year old Amazon indigenous person. A team of researchers from Brazilian dental colleges went deep into the rain forest to see if they could link a person's age to teeth wear. They found that the correlation could be made in the rain forest village but not in urban settings. This person has the same square pattern found in ancient jaws. Like those jaws, the canine teeth are in line with the incisors in the front of the mouth and it appears that tooth wear is on the lip side of the front teeth.

The numbers in the photo report the degree of wear. Below are brief quotes from the study.

" A strong relationship between tooth wear and chronological age in the three indigenous groups was observed.... For the control urban population group, no statistically significant association was found between tooth wear and chronological age."

"No significant differences were observed between males and females from the study sample; these two groups were therefore combined."

"Teeth are often used as "tools" in traditional communities. Expressions such as "using teeth as tools" and "a third hand" have been used to describe prehistoric reconstructions of human behavior. "

A large number of articles have been written about the size of Neanderthal's nasal opening. Various theories have been explored to explain why it is so large. Perhaps it is because of the need to warm frigid air with extra large concha. Or due to the need to of Neanderthals to take in great quantities of air to support their highly active lifestyle. One line of articles estimated a large caloric intake and speculated that the large nasal apparatus was needed to get enough air to burn the calories. Computer studies with complex statical analysis have

The flat front of the jaw and worn teeth are likely tied to preparing hides for clothing and domestic use. Methods might differ which would cause varying teeth wear patterns. With Inuits and Native Americans it is women who prepare hides. In both cases, teeth are used to work and soften the hides. There may have been a division of labor between the sexes in archaic humans. Researchers have found differing groove patterns between male and female Neanderthals.

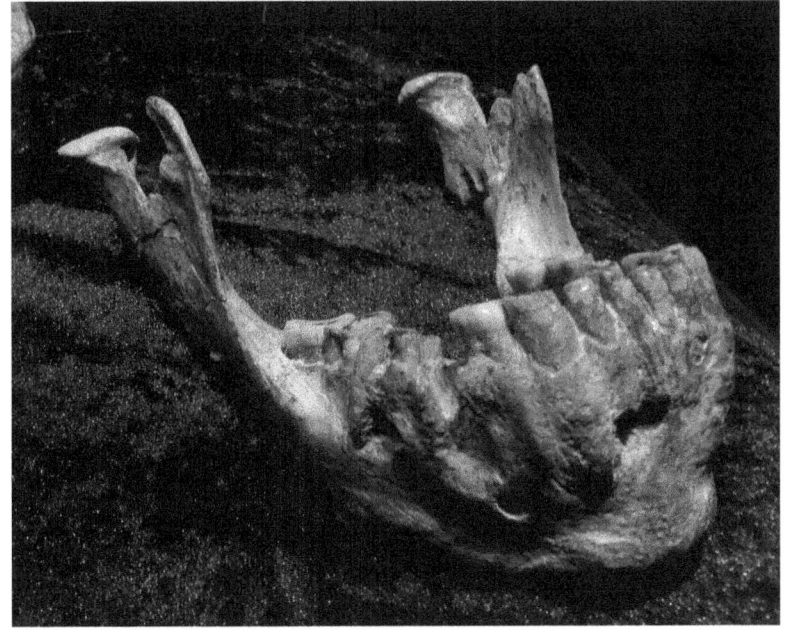

This Homo Erectus jaw is from Dmanisi, Georgia and is dated at between 1.85 and 1.77 million years ago. It is the oldest hominin fossil in Eurasia. The entire skull shown on pages 40 and 92 it has a very robust facial structure and jaw indicating heavy chewing forces. It has the same squared pattern and worn teeth of the later jaws. The buttress lines from the two canines come together in what appears to be a chin. In our short jaws, chins are prominent but gorillas and other primates have them too, just farther back on the jaw. As in this example, chins are buttress supports for the canine teeth.

One of the things we can learn from these examples is that none of these people could bite their fingernails. I am decades older that the Amazon indigenous person and likely the rest of the examples but my incisors are still sharp. They traded their ability to cut with their incisors to do other work. Was it the same work through the ages?

been undertaken to prove, or disprove, or the usual default, do a great deal of scholarship but not take a stance.

Previously, I wrote about noses being organs of smell and not primarily for breathing. If I am correct then the answer does not lie in efficient breathing or warming breath except to aid smell. Neanderthal's wide nasal opening is linked to the bio-mechanical architecture of skulls.

It seems likely that the Amazonian example is not an inherited trait but has come about by using the mouth to do work. It would be interesting to know what this person was doing. If other villagers had normally arched jaws, and whether related people in urban centers had this square configuration or the normal arch form.

Answers to these questions would indicate if the ancient jaws were formed by heredity or usage and what type of activity caused the change. The studies of Neanderthals explored high biting force as the cause of the their facial changes. I suspect this living person constantly used the mouth for a tool but not necessarily needing great biting force. The changed configuration may have come about by constant low grade stress.

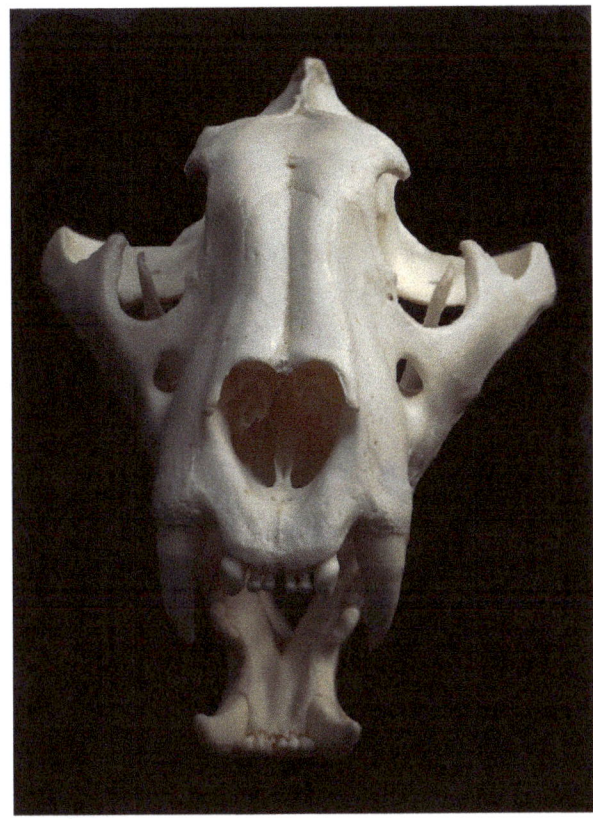

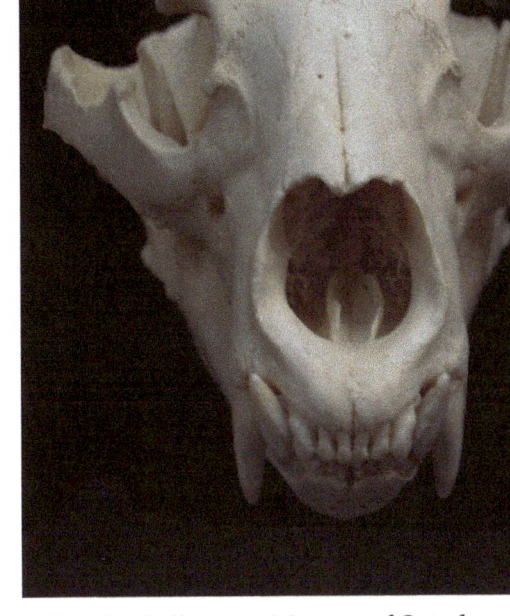

Lion skull   *Museum of Osteology*   *Grizzly skull*   *Museum of Osteology*

It perhaps seems strange to introduce a lion and a grizzle bear skull into the conversation about primate heads but these skulls follow from a common mammalian ancestor. Although modified, they use many of the same mechanics. These are carnivores and their mouths and fangs are designed for killing and eating prey. Primates have four incisors but these carnivores have six. The two outside incisors are more like small fangs and between them and the upper large fang, they provide shearing surfaces for each side of the lower fangs.

Both the bear and the lion are using the margins of the nasal opening to carry dental forces from the center and two outside incisors. The edge of the nasal opening in the bear is clearly a structural member for both the fang buttress and the outside fang-like incisors. In both skulls the buttress root for the fangs is along the outside edge of the nasal opening.

The skulls both have wide nasal processes to contain the dental forces and also carry them up into the brain case. We primates have narrow noses having been restricted in our masticatory biomechanics by our tarsier-like ancestor. Still, our noses do carry some pressure which causes the double arch eyebrow ridges in Neanderthals and some other hominins.

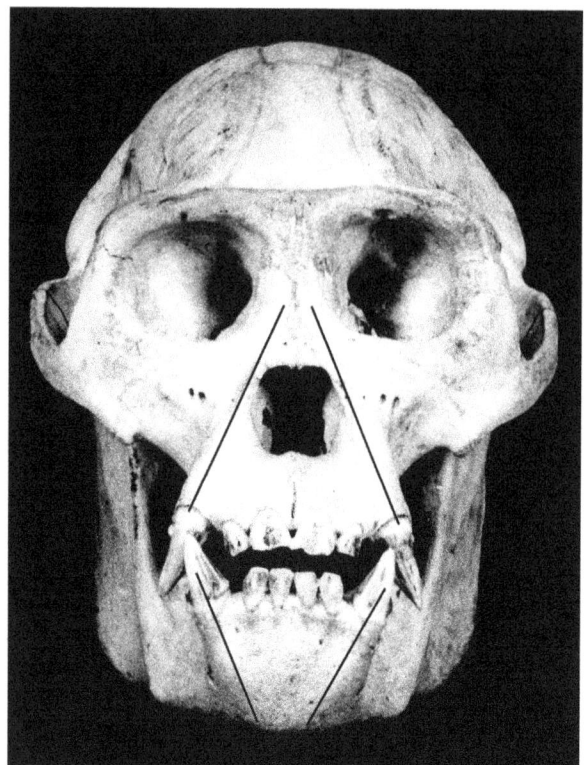

Gorilla Growth Series skull with lines added to show the upper and lower canine buttress supports. Note the gorilla has a chin which buttresses the lower canines but it is further back on the jaw.

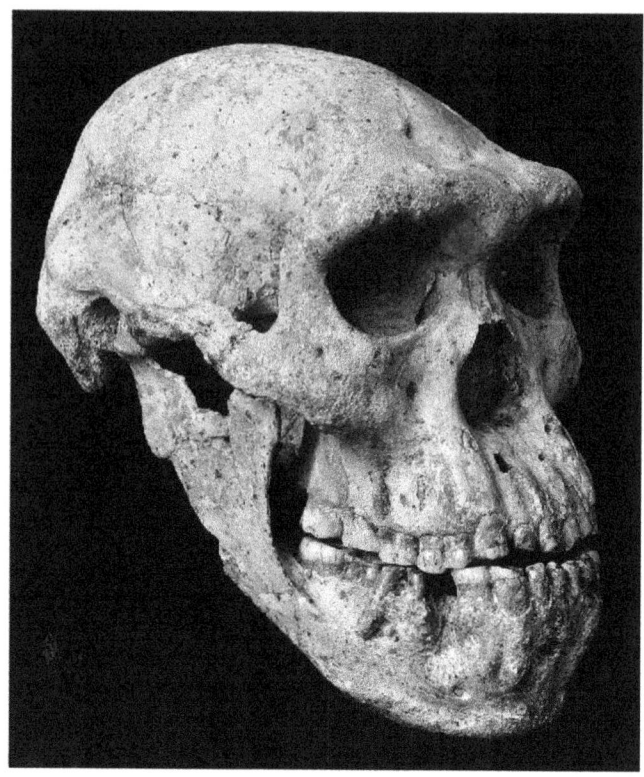

Skull 5 from Dmanisi, Georgia. Image credit: Guram Bumbiashvili / Georgian National Museum. This image is in its original Franfurt Horizontal position. The buttress line from the lower canine points to a small chin.

Like in the lion and bear, the buttress root of the canines in apes is associated with the outside edge of the nasal opening. In both the gorilla and the Dmanisi skulls, the outside edge of the nasal cavity supports canine pressures. The Dmanisi jaw, like the Neanderthal and Heidelbergensis jaws, has a flattened arch that brought the canines forward and also pushed them outward. The association of the canine root with the edge of the nasal cavity has the effect of widening the nasal cavity which, in case of Neanderthals, has launched a thousand papers.

Dmanisi skull indicates extraordinary dental pressures in the depth of the maxillary, the lower jaw, and in the robust zagomatic process and arch. These are more robust than in the senior gorilla's. While the gorilla's front teeth splay out, Dmanisi's are in a more powerful vertical angle. The gorilla has a chin to buttress the lower canines but it is further back on the jaw. Human chins serves the same purpose but due to our short jaws, protrude which is analogous to eyebrow ridges on other primates.

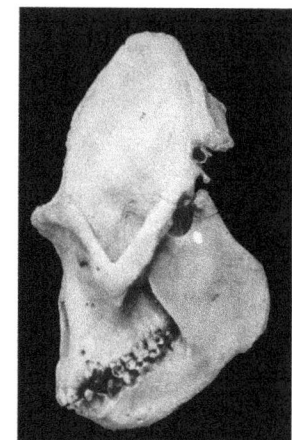

RECAP

Life's process of natural design has been able to achieve enormous variety and complexity while working within tight restraints and using quite limited methods. Its great advantages have been time, close to four billion years, and continuity.

The inheritance code that each of us carry in every cell of our body has managed since the dawn of divisible life, a period of nearly four billion years, to avoid dying. During this immense span of time, our personal code has never ceased to exist. It has always been able to construct an organism that stayed alive long enough to reproduce. Like the Hindu trinity of Brahma the creator, Vishnu the maintainer and Shiva, the destroyer, the code's interests are served in creating a new individual and then building and maintaining that individual until the period of reproduction is passed. The code then programs the demise of the individual to clear the way for the next generation where, hopefully, it now lives. Our genetic code has always managed to avoid dying by jumping ship to the next generation.

The genetic code always needs a living organism to carry it into the future. It is hard to grasp that the organization of chemicals that programed the first primitive organisms is the same one that has made us. We encompass all our ancestors from single celled organisms to our present astonishing complexity. For almost four billion years our inheritance code has been adapting by self programing.

The history of life has the pattern we'd expect from a self-programing process; very slow at first and then a steeply ascending developmental curve. In life's nearly four billion years approximately three quarters of that time had only relatively primitive bacteria. A billion years ago, single celled eukaryotic organisms evolved and they later learned how to cooperate to form multi celled organisms. Only in the last six to seven hundred million years did the spectrum of multicellular eukaryotic life evolve.

We are limited in understanding evolution by the structure of our consciousness which is programmed to see the world being made up of a myriad of separate entities. In thinking about evolution, we run into the problem of defining the entity. Where is the thing of evolution? Is it the organism, the species, or the process? None of the stages between the general process and the specific organism are totally satisfactory. Natural design is not a thing, it is like a river, a flow. This flow uses the organisms that it designs and builds as vehicles to carry it along. As it has moved along through one organism after another, it has learned by slow accumulation to control the staggeringly complex metabolic inter-relationships within organisms that are necessary to maintain life. Life's goal is not the organism or the species but simply its own survival. It can only survive by adjusting to ever-changing reality. From life's perspective, a worm is as good as an Einstein.

Life has learned to tune and modify its organisms to new conditions by adjusting their physical, behavioral, and metabolic interrelationships. It will reform its particular organic line to take it towards a source of substance and survival. Like us, it sometimes run up a dead end chasing a meal. A three pound toy poodle, for instance, is a wolf making a living being cute. I don't ascribe consciousness to this process. Thought and consciousness are not the same thing but even the word "thought" is likely overblown in this context. Like a computer, it runs programs.

In nature, form does follow function. Bones flow and change form to maintain relation-ships. A major impulse of the natural system is to remove any extraneous materials to save energy. Designers refer to this as "economy of means". Bone grows along lines of stress and that bone form will fade when stress is lessened or absent. When astronauts become weightless their bodies immediately start to flush out calcium from their bones. The processes that monitor their bodies suddenly finds that it has no need for bones. The system has never encountered weightlessness before in its billions of years but it has encountered a lack of stress and responds accordingly.

Multiple processes monitor and adjust a myriad of relationships within our bodies. Organ-isms, even the simplest ones, are essentially chemical computers. External conditions will engage certain programs. Heat or cold cause shifts in metabolism and engage cooling mechanisms. Constant exercise over time will promote blood vessel growth to deliver an enhanced blood flow to the muscles. The organism's computer adjust for internal or exter-nal changes, like the sudden loss of gravity.

Life's process does not allow its organisms to remain for too long in a stressed condition. Stress, continuous uncomfortable conditions, and the resulting inefficiencies are bad for the code's own survival. The design process will cure systemic stress by modifying its or-ganism's design. We humans insisted on moving around on our hind legs because it gave us an advantage. The design of our heads and bodies adjusted to be able to function in this vertical mode without systemic stress. In other words, it reacted to feedback. I dem-onstrated this in my analysis of primate head design, a feedback interaction between the natural design program and the real world experiences of the group. The process is evident even within the life cycle of individuals, both ape and human.

Numerous moments of divergence, what we could call decisions, were made along the way and they became part of the fabric. Once in the fabric, they can only be maintained, modified, or suppressed but not deleted. Mammal heads are improvisations on a theme sharing basic mechanics and structures from a distant common ancestor. Primate heads are a subset sharing some unique attributes from a more recent common ancestor.

Evolution doesn't know the future and relies solely on modifying what already exists. Unlike human engineering, it can't go back and start over when things get wonky. Problems arise when evolution's early solutions lead into a dead end. Nature can't develop anything from crashed, extinct species but can only build from those that persist. Its only option is to develop a compromise to get around the problem. Animal heads are one of those compromises, and an ugly one at that. A single sentence will negate all the arguments proposed by religious Intelligent Designers. **It cannot plan for the future and its structures are often not good design.**

No one should have a head or a neck. Just think of how predators throttle their prey, or hangings, our ubiquitous use of helmets and on and on. Early multicellular organisms articulated a system for nerve communication from already existing chemical communications being used within and between cells. The result is a slow chemically based nervous system. Hundreds of millions of years later, we still depend on a chemical transfer at the synapses.

Heads became necessary when creatures needed more speed and behavioral sophistication to interact with their external world. A central decision making unit, the brain, was placed at the front with the primary senses plugged directly into it. For the same reason the head is also the port of entry for food, water, and air which are monitored for suitability before being ingested. Our lives have two centers, our head and our body, connected by the fragile umbilical of the neck.

Nerve impulses travel at a speed of between two and two hundred meters per second. In that second, an electronic signal would be two-thirds of the way to the moon. A creature with an electronic brain and nervous system would not need a head. Its brain could be at its center, well nourished, and protected and not, as in the case of Marie Antoinette, subjected to being misplaced. This creature's sensory apparatus would be away from the brain and dispersed around its periphery. It strikes me as a bit silly that engineers and designers keep putting heads on their robots.

The laboratory skull that sat on my drawing table for the summer taught me a number of things. I was only interested in it as a functional object. I learned a few of the scientific names for parts of the skull but not many. The bone plates, their boundaries, and their Latin-based nomenclature didn't help my understanding very much. Where such boundaries occur they are often in the middle of a complete form that has a function.

The human skull, I found, has only certain places where its form can be modified. As we often discover about our own lives, conflicting needs will stifle change. There are configurations where change is prohibited and places where the design is fluid. Once you understand the game, it is not too hard to predict the outcome.

Evolution is a conservative process and it keeps its working configurations unless they are challenged by reality. When challenged, evolution can only modify existing structures such as turning scales into feathers and hair or suppressing others like hen's teeth and fur on humans. Using the image of an onion as a metaphor, the inner core layers are old decisions which are immutable. All aspects of an organism have core immutable layers and outside layers that can be modified. The adjustable outside layers are not just the fir and feather, but all the organs and relationships within the organism. These can be adjusted but not deleted.

Image fallacies have been very subversive in the study of human evolution. We all, tribal hunter/gathers through Ph.D.'s, start with the same set of concepts of the figure and being a trained anatomist doesn't seem to help much. Next time you are in your doctor's office, look at the anatomical chart on the wall with all the body's muscles accurately displayed. In many of these charts, just adding skin and some tribal ornaments to this scientific figure, and it could easily pass as an idol in a tribal village.

BIBLIOGRAPHY

An Introduction to Human Evolutionary Anatomy
Leslie Aiello and Christopher Dean, Academic Press 1990

Anatomy for the Artist
Jeno Barcsay, Barns and Noble Books 1999

Edward S. Curtis Chiefs and Warriors
Christopher Cardozo Callaway Editions, Bulfinch Press 1996

Leonardo Da Vinci, The Anatomy of Man
Martin Clayton, Little Brown and Company 1992

The Illustrated Origin of Species,
Charles Darwin, Abridged and Introduced by Richard E Leakey, Hill and Wang 1979

The Anatomy of Nature
Andreas Feininger, Crown Publishing 1956

Fossils of All Ages
Jean Claude Fischer Yvette Gayard-Valy, Grosset & Dunlap 1976

Hen's Teeth and Horse's Toes
Stephen Jay Gould, W.W. Norton & Company, 1983

From Lucy to Language
Donald Johnson and Blake Edgar, Simon and Schuster 2006

The Cambridge Encyclopedia of Human Evolution
Steve Jones, Robert Martin, David Pilbeam, Sarah Bunney editors 1992

Human Origins The Fossil Record
Clark Spencer Larsen, Robert M.matter, Daniel L. Gebo, Waveland Press 1998

The Evolution of the Human Head
Daniel E. Lieberman, Harvard University Press 2011

Evolutionary Biology of the Primates Osman Hill, Academic Press, London New York 1972,

The Story of the Human Body
Daniel E. Lieberman,Vantage Books 2013

The Descent of Woman
Elaine Morgan Souvenir press 1972

Atlas of Human Anatomy for the Artist
Stephen Rogers Peck, Oxford University Press 1951

Evolution A view from the 21st Century
James A. Shapiro, FT Press Science 2011

Extinct Humans
Ian Tattersall, Jeffrey Schwartz, Nevraumont Publishing Company 2001

The Monkey in the Mirror
Ian Tattersall Harcourt Inc. 2002

Skulls
Simon Winchester, Black Dog and Leventhal Publishers, 2012

For two to three years now I have been getting almost daily papers from Academia they now number in the hundreds. Originally I was seeking information on Neanderthals but the subjects have widened in scope. Below is a brief smattering of the papers.

"Gunz, P. et al. (2012) A uniquely modern human pattern of endocranial development. Insights from a new cranial reconstruction of the Neandertal newborn from Mezmaiskaya. Journal of Human Evolution. 62(2):300-313." by Jean-Jacques Hublin

"Neubauer, S., Gunz, P. and J.-J. Hublin (2009) The pattern of endocranial ontogenetic shape changes in humans. Journal of Anatomy. 215(3):240-255." by Jean-Jacques Hublin

"Correlation of cranial and mandibular prognathism in extant and fossil hominids" by Louise Leakey 2005, Transactions of The Royal Society of South Africa

"Isotopic evidence of early hominin diets" by Matt Sponheimer 2013, Proceedings of the National Academy of Sciences of the United States of America

"Craniofacial modularity, character analysis, and the evolution of the premaxilla in early African hominins" by Brian Villmoare

""Out of Africa I": Current Problems and Future Prospects" by Robin Dennell 2000

"Size, shape, and asymmetry in fossil hominins: the status of the LB1 cranium based on 3D morphometric analyses" by Karen Baab

"Human Life History in Primate Perspective: 2009 AAPA, Chicago" by Barry Bogin

"Long-term patterns of body mass and stature evolution within the hominin lineage" by Manuel Will

"Implications of KSD-VP-1/1 for early hominin paleobiology and insights into the last common ancestor (LCA)" by Marc Meyer

"On the Variability of the Dmanisi Mandibles " by Mark J. Sier

www.ingramcontent.com/pod-product-compliance
Lightning Source LLC
Chambersburg PA
CBHW052142170526
45159CB00017B/3136